The Untold Story of Everything Digital

Bright Boys, Revisited

by
Tom Green

CRC Press
Taylor & Francis Group
Boca Raton London New York

CRC Press is an imprint of the
Taylor & Francis Group, an **informa** business

A CHAPMAN & HALL BOOK

CRC Press
Taylor & Francis Group
6000 Broken Sound Parkway NW, Suite 300
Boca Raton, FL 33487-2742

Printed on acid-free paper

International Standard Book Number-13: 978-0-367-35560-9 (Hardback)
978-0-367-22007-5 (Paperback)

Library of Congress Cataloging-in-Publication Data

Names: Green, Thomas J., author. Title: The untold story of everything digital : Bright boys revisited / by Thomas Green. Other titles: Bright boys Description: Special, anniversary edition. | Boca Raton : CRC Press, 2019. | Includes bibliographical references and index. | Summary: "The Untold Story of Everything Digital: Bright Boys, Revisited celebrates the 70th anniversary (1949-2019) of the world "going digital" for the very first time-real-time digital computing's genesis story. That genesis story is taken from the 2010 edition of Bright Boys: The Making of Information Technology-1938-1958, and substantially expanded upon for this special, anniversary edition. Please join us for the incredible adventure that is The Untold Story of Everything Digital, when a band of misfit engineers, led by MIT's Jay Forrester and Bob Everett, birthed the digital revolution. The bright boys were the first to imagine an electronic landscape of computing machines and digital networks, and the first to blaze its high-tech trails"-- Provided by publisher. Identifiers: LCCN 2019024850 (print) | LCCN 2019024851 (ebook) | ISBN 9780367220075 (paperback) | ISBN 9780367355609 (hardback) | ISBN 9780429340246 (ebook) Subjects: LCSH: Air defenses--United States--History. | Electronics in military engineering--United States--History. | Electronic digital computers--History. | Information technology--History. | Computer engineers--United States--Biography. Classification: LCC UG730 .G74 2019 (print) | LCC UG730 (ebook) | DDC 623/.043097309045--dc23 LC record available at https://lccn.loc.gov/2019024850 LC ebook record available at https://lccn.loc.gov/2019024851

Visit the Taylor & Francis Web site at
http://www.taylorandfrancis.com

and the CRC Press Web site at
http://www.crcpress.com

Printed and bound by CPI Group (UK) Ltd, Croydon, CR0 4YY

Contents

Author's Note

The Untold Story of Everything Digital: Bright Boys, Revisited celebrates the 70th anniversary of the world "going digital" for the very first time (1949–2019). To wit, it is real-time digital computing's genesis story. That genesis story is taken from the 2010 edition of *Bright Boys: The Making of Information Technology, 1938–1958*, and substantially expanded upon for this special, anniversary edition.

The books are a complementary pair. It's recommended for readers looking to examine the more expansive quest for Information Technology (1938–1958) to see *Bright Boys*, especially for its bibliography and footnotes that cover both books in depth.

Foreword

This book tells the story of the Barta Building and the external world from which the Whirlwind computer emerged. It is an authoritative account about a team of individuals who pioneered technology that forever changed our world. But there is more to the story—the internal MIT environment that fostered innovation. Especially in the 1940s, MIT can be described as a free enterprise society in which one could do anything if it was honorable and one could raise the necessary money from the outside. In this freedom of action, I believe MIT differed from almost all other academic institutions.

Such freedom to innovate resided in the Division of Industrial Cooperation (DIC) directed by Nathaniel Sage. The DIC had been set up in the 1920s to handle contracts and funds, mostly from industrial companies that wanted to sponsor specific research projects. The DIC then became the contracting vehicle for government-sponsored activities during and after World War II. The DIC was parallel to and mostly independent of the MIT academic departments and could independently hire people.

Nat Sage set the tone of the Division of Industrial Cooperation. He favored results over the restraints of contracts and bureaucracy. As the son of an Army officer, he grew up in Army camps around the world. Somewhere in that experience, Sage became a very good and self-confident judge of people. There were those at MIT whom he trusted implicitly, and those whom he would not trust farther than he could watch them. Sage trusted Gordon Brown, my mentor, and Stark Draper, famous for gyro computing gun sights in World War II and the guidance system for the moon landing. I was also on his favored list, and he was a tower of strength in tough times, of which we had many.

An incident in the later stages of designing the SAGE computers showed Nat Sage's way of dealing on our behalf with distractions. We had several engineers who had spent a stressful week in Poughkeepsie, New York pushing the IBM engineers to design for the reliability to which we were committed. Came Friday afternoon, they wanted to be home for the weekend, and rather than face the tedious trip to New York and a change of stations for a train to Boston, they chartered a plane to take them home.

When the Air Force finance officer saw the charge for the plane, he came to Nat Sage to berate us and demand some kind of punishment. Sage was sympathetic, said he would look into it, and after the officer left, he put the papers in his desk. Several months later he mentioned the officer's visit to me.

Unlike most organizations, in the DIC bad news flowed upward. If there was no news, those at the upper levels assumed matters were under control. The upper levels were there to assist when there was trouble, and when necessary we took problems "upstairs" for help. Also, administrative levels were not barriers to communication; everyone was free to skip levels, both up and down, to go directly to the place where discussion could be most effective.

The core team that created Whirlwind emerged from the early 1940s in the MIT Servomechanisms Laboratory directed by Gordon S. Brown. Brown was principally responsible for my career at MIT. Brown ran what at the time seemed like a chaotic organization. World War II projects came in; he turned most over to graduate students to manage, keeping a few projects for himself to manage. We, as employees of the boss, had to compete with him for resources and influence. Brown was impatient, with little sympathy for incompetence.

He would "walk on" those who did not defend their positions, but respected and gave tremendous help to those who stood up to him and demonstrated justifiable independence. I think my turning point with him came one day when I had an issue to discuss with him and went to his office door where he was sitting at a desk facing the door. I stood in the door waiting to be invited in and was sure he knew I was there.

After a long time, he looked up and said, "Get out!" At that point I walked to his desk, sat on the paper he was writing, and told him I would not leave until he paid attention. It was a rough environment but powerful in character building.

In developing military equipment, we gained experience in the total sequence of new products—defining the objective, innovating new ideas, studying the theory of operation, making demonstration models in the machine shop, designing for production, devising tools for factory manufacture, solving factory production problems when they arose, and going to the military field to analyze and correct failures in operation.

When we started such work, MIT was without guards and military security; to fill in, I was given a City of Cambridge police badge and a pistol permit to double as a guard. At age 24, I signed drawings in the space reserved for the Chief of Army Ordnance and sent them directly to factories; such was the freedom and urgency of wartime. Here we came to a deep and abiding faith in Murphy's Law that, if anything can go wrong, it will. That enduring belief served us well as we embarked on designing reliable computers. The Servomechanisms Laboratory under Gordon Brown trained innovators.

Today, many critics lament the lack of innovation in our society and draw the conclusion that more emphasis on teaching mathematics and science will lead to innovation. That will probably fail. Pressures in the present school systems suppress innovation. Innovation comes from repeated successes in innovating. Innovation means trying ideas outside of the accepted pattern. It means providing the opportunity to fail as a learning experience rather than as an embarrassment. It means living part of the time outside of the traditionally accepted track.

An innovative spirit requires years for developing the courage to be different and calibrating oneself to identify the effective region for innovation that lies between the mundane and the impossible. Almost none of the conditions for developing innovative attitudes are to be found in today's K–12 education. In fact, the traditional school powerfully suppresses any tendency toward being innovative. Both teachers and students are driven to conform.

However, today's educational shortcomings did not exist in Brown's laboratory. Each project opened into uncharted territory. Each provided a wide range of experience that far exceeded that of most people in industry, government, and academia. It was this wealth of experience and building the personal character for innovation and entrepreneurship that produced the team that was capable of pioneering the digital frontier.

Gordon Brown had an even more crucial role by launching the activity that led to the Whirlwind and SAGE programs. The end of World War II terminated the military projects in the Servomechanisms Laboratory. I assumed that I would be leaving MIT either to find an industrial position or to start a new company to carry forward and apply our feedback system knowledge to physical engineering applications. Then, Gordon called me to his office to ask if I would like to stay and choose one of the ten projects that he had on a list.

I decided to pick one from the list. Captain Louis [sic] de Florez at the Navy's Special Devices Center in Port Washington, Long Island, had proposed it. De Florez was an uninhibited innovator. One story from his early career comes from when he worked for an oil company in Texas. Refineries were threatened by fire and explosion when a pipe eroded from the inside until the wall was weakened, and a major bursting and explosion could occur. De Florez devised a warning system by drilling small holes part way into the pipe walls. When the inside eroded to the bottom of a hole, oil would spray out as a warning. De Florez was a commanding figure with a pointed waxed mustache.

Also, as far as I know, he was the only person able to get permission to land a seaplane in the Charles River sailing basin in front of MIT. I was there once when the Metropolitan District Commission cleared the basin of sailboats; de Florez landed his plane and came to the MIT alumni reunion. After lunch, when the speeches were becoming dull, he left in a roar from his water takeoff that overwhelmed the loud speakers at the reunion. The de Florez project was to go beyond the existing flight trainers for pilots, which were tailored to the characteristics of a known airplane. The project was to create a cockpit that would exhibit the characteristics of a proposed plane, based on wind tunnel data of a model of the plane. Thus started the sequence described in this book.

Although the Division of Industrial Cooperation was outside of and parallel to the MIT academic departments, projects did have a tenuous connection to a relevant department. Almost always, there was a faculty member who bridged between a DIC project and an academic department. Gordon Brown played the role of connecting the digital computer activity to the MIT Electrical Engineering Department.

We did have frictions in trying to awaken a traditional electrical engineering department, devoted to power generation, to the arrival of computers. At one point we presented, with only partial success, a lecture trying to convince the electrical engineering faculty that it was not only possible but also desirable to use binary arithmetic for computation.

Our relationship to the Electrical Engineering Department gave me access to incoming graduate student applicants to become full-time research assistants. They were allowed to take two subjects per term toward their degree. By reviewing applications myself, I effectively had first choice of those applying to MIT. We accumulated an impressive group. Many graduate applicants were coming back from military service with a maturity and real-world experience that is unusual for students. They were ideal recruits for the coming information age. Their MSc theses on aspects of computers qualified many to present papers at the national and international technical conventions.

Because of the continuing budget pressures in the early phases of the Whirlwind computer, we were subjected to annual reviews by panels appointed by the Office of Naval Research. Most of these were a distraction from our mission, but occasionally a point was raised that improved our thinking. An especially important advance came when ONR appointed a mathematics professor, Francis Murray of Columbia University, to be inquisitor-of-the-year. Murray came on a Saturday and we met in my office.

At one point he asked, "What are you going to do about the electronic components that are drifting gradually and are on the edge of causing mistakes? Any little random fluctuation in power, or streetcars going by, will cause circuits to sometimes work and sometimes not." This was a very perceptive and powerful question. Inexplicably, we had done nothing about it. It was such a pointed question, and obviously such an important one, that I felt an immediate answer was essential. I said to him, "Well, we could lower the voltage on the screen grids of tubes to change their gain and convert behavior from a marginal to a permanent failure and then it would be easy to find."

He thought it was a good solution and so did we. The next week we started designing such capability into Whirlwind. The "marginal checking" system in Whirlwind carried over into the SAGE Air Defense system, adding another factor of ten to the reliability.

During the design of Whirlwind, we were strongly criticized for the amount of money we were spending to develop entirely new circuits and devices that would be reliable. The total Whirlwind cost over some seven years was $4.5 million; it would not be long after until individual production line machines were costing that much.

The matter of cost was one of the things that the outside world understood least. Whirlwind was being judged in the context of mathematical research, in which the salary of a professor and a research assistant was the standard by which projects were measured. We were spending way beyond that level and were seen as running a "gold-plated operation." Although the gold plating might occasionally have been excessive, in retrospect, I think there was reason for it.

An organization has great difficulty maintaining two contradictory standards. If you're going to have high performance and high quality in the things that matter, it is very difficult to have low quality and low performance in the things that, perhaps, don't matter. The two standards bleed into one another to the detriment of both.

For example, at an early demonstration for important people, we didn't want them sticking their fingers into the high voltages in the circuit racks of Whirlwind. I asked somebody to get rope to put along the aisles so visitors wouldn't walk among the vacuum tubes. A nice-looking white nylon rope was procured and installed. During the demonstration, I saw some of our critics fingering this beautiful rope and looking at one another knowingly as if to say, "That's what you would expect here." It may not have cost any more than hemp rope, but it reinforced that impression of an extravagant operation.

Another example was the Cape Cod display scopes built into cabinets made of mahogany-faced plywood. Although our cabinetmaker made these quite inexpensively, people looking at those mahogany cabinets were reinforced in their thinking that we were extravagant. Eventually we solved this problem by spending additional money and painting the cabinets gray.

We came out of the Servomechanisms Laboratory with a central group that understood the full potential of teamwork. Of course, later, the team unity was reinforced as we stood together on the computer frontier to repel critics. There was complete sharing of information. In a bi-weekly report, distributed to everyone, each person reported on progress and difficulties. There was little jockeying for personal advantage.

Every person has strengths and weaknesses. A team must have a shared vision of the future, a sensitivity to political matters, the capability of developing people, technical competence, the courage to transcend adversity, salesmanship, integrity, the ability to put long-range goals ahead of the short term, and a shared understanding of the individual strengths and weaknesses within the group. We had those characteristics well represented, scattered throughout our group. No person had all these skills. For every person there would be a glaring hole in one or more of those dimensions.

Yet, it was a group that understood each other well enough to use people in situations where their strengths prevailed and have others compensate for their weaknesses. Out of that came an organization that was able to be much more effective than most of those we see around us in technology and corporations at the present time. It was an organization possessing power based on a clear vision and consistency that could dominate the military and corporate structures that grew around the North American Air Defense system.

Jay W. Forrester

Author

Tom Green, journalist, writer, video & TV producer, has been reporting and producing content on technology for over two decades.

Green is the founder (2017), publisher, and editor in chief of *Asian Robotics Review.*

Previously, 2012–2016, he launched and was founding editor in chief of *Robotics Business Review* (a property of EH Publishing). Green was also on-air host and lead researcher (2013–2016) for Robotics Business Review's webcast programs, as well as lead editor and contributing author for the publication's annual series of robotics research reports.

Green has spoken at national and international robotics events and conferences, and has been the subject of interviews on robotics with Barron's, The *Wall Street Journal, Swissquote,* and *CNN Money.*

Formerly, as a TV writer/producer at Boston's ABC affiliate WCVB-TV, Channel 5, he wrote and produced the news program *Lifelines,* and was a lead writer on the weekly sitcom *Park Street Under.* His work has been twice nominated for Emmy Awards.

His stage plays were produced at Boston's Next Move Theatre and then reproduced as radio plays for National Public Radio. Green's "In the Room the Women Come and Go" won a UNICO National Short Story Award.

Green owned and operated his own video production company for 10 years, producing video for corporations, broadcast, and cable TV.

He is the author of the Amazon best seller *Bright Boys: The Making of Information Technology: 1938–1958* (Taylor & Francis Group 2010).

Green's newest book, *The Untold Story of Everything Digital: Bright Boys, Revisited* (Taylor & Francis Group), was published in September 2019.

His mystery/detective novel, *The Flowered Box* (Beaufort Books & PaperJacks 1990), a novel of suspense, was set in Boston; in 2020, a reissue of *The Flowered Box* will be reset in Bangkok.

Green's science fiction novel "*S*" is scheduled for release in 2021.

For 20 years, Green and his wife (she's Thai) have spent much of their time in Asia, where they have friends, family, and a home.

Introduction

If the Barta Building Could Speak
The Most Important Building in the History of Digital Computing

Exterior of Barta Building, 211 Massachusetts Avenue, circa 1949. (Courtesy of MITRE Corporation.)

> *We were engaged in an endeavor that no one else could claim to understand.*
>
> —Jay Forrester

The Beginning That Changed Everything
Measured by the technology pioneered within the confines of its old brick walls, the Barta Building, just like Dr. Who's TARDIS, is bigger on the inside than it is on the outside. Our digital universe was born there over a remarkably brief period of time—1948–1953—and because of that discovery, the provenance of every digital device has but one address: 211 Massachusetts Avenue, Cambridge, Massachusetts.

The Untold Story of Everything Digital: Bright Boys, Revisited is the story of how our high-tech world came to be. It's digital technology's genesis story as lived through the pioneering events of a small group of people who made it happen. Amazingly, it all took place in a single location, the Barta Building, an old, former commercial laundry a block or so away from the front door of the Massachusetts Institute of Technology.

Within the Barta Building, occupying some 2500 square feet of floor space, was a monstrous, Rube Goldberg-looking machine called the Whirlwind computer (1948–1959). Although the exact eureka moment of Whirlwind's first digital ping is shrouded in a bit of mystery, by the fall of 1949, it was definitely purring away, solving simple equations in real time and displaying answers on an oscilloscope. It was a first-ever for real-time digital computing.

Although what was birthed in the Barta Building has transformed just about everyone and everything on planet Earth, the building has yet to make it to the National Register of Historic Places—those "deemed worthy of preservation for their historical significance." Recognition has been scant. The best that the old place has mustered is an IEEE commemorative plaque at the front door, placed there in 2012. For decades, it was known on MIT campus maps simply as building N42. And doubtless, none of the thousands of students and staff who traipsed through its corridors over the years, probably all toting some sort of digital device, ever realized that maybe a respectful bow at the threshold was in order. Who knew?

Recognition for the engineers who designed and built Whirlwind fared not much better. Forty years after the fact, in 1989, the digital leadership duo of Jay Forrester and Bob Everett were awarded the National Medal of Technology. The others were content just to be happy for Jay and Bob.

What they did together inside the Barta Building is what galvanized a decade of their lives as engineers and bound them together as friends and comrades for life. As one of them, Bob Wieser, mused years later, "Sometimes I ask myself why this was such an interesting experience, the like of which I haven't had since."

If the Barta Building could speak, oh the magical tale it could tell about the digital beginnings that changed everything. It would tell the story of 30-year-old Jay Forrester leading a team of cocky but brilliant engineering bright boys on an amazing journey that in a single, white-hot decade (1946–1956) of fantastical invention made the world go digital ... for the very first time.

Young, unknown, with zero standing in academic circles, they were frequently discredited as brash, wasteful "boondogglers" with crazy ideas that they thought would change the world for the better. Such adversity made them struggle mightily and unnecessarily, but they persevered ... and won.

The Barta Building's band of misfit engineers birthed the digital revolution. They were the first to imagine an electronic landscape of computing machines and digital networks, and the first to blaze its high-tech trails.

Their work in the Barta Building created a profusion of technological firsts that is as yet unmatched in any single engineering project since.

A list of just the top ten inventions witnessed by the Barta Building, from a list that stretches on for dozens more, readily displays the bright boys' brand of rapid innovation and talent for discovery that now form the bedrock of our digital age:

Top Ten Barta Building First-Evers

#1. World's first-ever high-speed, real-time, electronic digital computer.

#2. First-ever software: Code warriors J. Halcombe Laning and Neal Zierler combined shorthand English and algebraic formulas for the first algebraic compiler. IBM's John Backus visited the Barta Building for a demo, then returned to IBM and invented FORTRAN (the first widely used high-level programming language).

#3. Jay Forrester invented magnetic-core memory (RAM): Magnetic-core memory made stable memory and the stored program possible for the first time. It became the computer industry standard for 20 years, until Intel's first chip in 1971.

#4. Impatient using punched paper tape, first-ever computer keyboard and monitor installed on Whirlwind.

#5. First-ever computer graphics program: Douglas Ross secretly developed a graphics program for Whirlwind and used his fingertip as a stylus while drawing directly on computer monitor. Predating Ivan Sutherland's *SketchPad* (ancestor of modern computer-aided design) later developed on "transistorized" version of Whirlwind, the TX-2, in 1963.

#6. Bob Everett designed and prototyped first-ever "light gun" in 1950 (precursor to the modern mouse). Touch the light gun to the monitor, and the computer responds.

#7. First-ever networking as data was transmitted via microwave from Bedford, Massachusetts, to Whirlwind in Cambridge, Massachusetts (20 miles distant). Proved to be too expensive; opted for telephone lines.

#8. First-ever modem (MOdulator—DEModulator): When AT&T protested that copper-wire telephone lines were unable to transmit data, John Harrington's team invented the modem.

#9. First-ever computerized air traffic control: Combined flight radar data and Whirlwind computer to organize and control air traffic. Became international standard for both military and civilian air transport.

#10. Dudley Buck invented the "cryotron," a superconducting switch which today is part of superconducting quantum bits (qubits) in quantum computing.

The list goes on; it's long and illustrious ... and mind-boggling!

Along the way, they poked a first-ever hole into the Infosphere and put on the first practical demonstration and use of information theory. Norbert Weiner and Claude Shannon had just given new, technical meaning to words like communication and message, writes Howard Rheingold in his *Tools for Thought*. "They demonstrated that everything from the random motions of subatomic particles to the behavior of electrical switching networks and the intelligibility of human speech is related in a way that can be expressed through certain basic mathematical equations."

John Harrington first put information theory into practice with the digital transmission of 1s and 0s as radar pictures over voice telephone lines.

Initially, the bright boys had little idea what exactly it was they dragged in through that gaping hole. Their search was for how best to manipulate digital information using their newly hatched creation, the Whirlwind computer. Information Technology, a term that wouldn't be coined until 1958, was an unexpected consequence, an eye-opening bonus.

Each time we boot up our laptops, step off an elevator, book an airplane ticket, get cash from an ATM, or even pop a steak into a microwave, there's an unseen thread that leads directly back to the bright boys' inner sanctum of discovery and innovation.

Of course, now everyone and everything is bound up in the clutches of this all-pervasive, all-powerful Information Technology, popularly known to most as just IT. IT is that high-tech panoply of computers, telecommunications, and the myriad devices that digitally connect and integrate information, equipment, and people.

Beginning with the bright boys' first information loom, built and first cranked into action in 1949 in the old laundry building, the two initials, IT, have subtly yet inexorably woven themselves into the fabric of our lives.

Today there are information looms everywhere, atop our desks, on our laps, and in our pockets and purses. Many of us spend most of our waking hours spinning information on them, creating, shaping, and forming information as needed, then sending it at the speed of light everywhere and anywhere at any time to anyone of our choosing.

We've gone so far as to define our times as an information age, an information economy, an information society, or an information revolution. Not so in the world that swirled around 211 Massachusetts Avenue. Back then there was a pristine simplicity to the notion of information. Even as late as the mid-1940s, information was meant mostly for the likes of the U.S. Census Bureau every 10 years, banks, and insurance actuaries.

As Thomas Haigh points out in *Inventing Information Systems*, there were, in all of the technical and scientific literature for 1946, only seven articles on information.

Today the word "information" has taken on a whole new spin: it's anything and everything ... and it's powerful. Charles Seif, the well-known physicist and journalist, likens us to information beings. "Each creature of the Earth," he writes in *Decoding the Universe*, "is a creature of information; information sits at the center of our cells, and information rattles around in our brains." The book's subtitle hammers home that conviction: *How the New Science of Information Is Explaining Everything in the Cosmos, from Our Brains to Black Holes.*

No one in 1946 could ever have fathomed how transmuted information would become when mixed in the crucible of technology.

Jay Forrester and his band of bright boys were as startled as any.

In the faint pre-dawn glow of the digital age—the mid- to late-1940s— electronic digital computers had much in common with Jurassic dinosaurs: large in body but very short on brainpower. Staggering behemoths of hardware, most were about the size of a gymnasium, yet with barely a fraction of the capability of a modern laptop. Hardly anyone gave much thought to an electronic digital computer doing anything more than munching huge quantities of numbers and then slowly regurgitating an answer. Hardly anyone, that is, except for a cocky bunch engineers holed up in the Barta Building.

The bright boys and their massive machine would, indeed, change the world.

Democratizing Computing

They changed the world by changing people, one by one, if necessary.

In an age when the few behemoth mainframes that existed were protected and horded by their builders, Forrester, Everett, and their band of engineers freely offered machine time on Whirlwind to anyone. Across the MIT campus and beyond, researchers from academia, business, and industry by the hundreds lined up at the digital oracle to get answers for their hardest of problems.

In evening classes, they even taught binary math and programming to the curious and to those looking to embark upon digital careers. They created true believers in their machine and in their technology. A diaspora of the "converted" grew into a large, ever-growing, and enormously influential fan club that proselytized the great machine far and wide.

During the entire decade of their pioneering the digital universe, they never wavered from giving back to those in need. Their machine always had the welcome mat out, any time day or night.

Edward R. Morrow called them "Unusual … and gifted."

The rest of us have yet to thank them for making the world go digital for the very first time.

Humble and Brilliant

Today, they are by and large anonymous, although their feats of engineering are an essential part of our daily lives. Not much happens today without the aid of digital electronics, and certainly little in the future will take place without it. Yet, the young engineers who made going digital possible for the rest of us are faceless, when they should be recognized and honored.

When I first met them as octogenarians and nonagenarians in 2003 (most all have passed since 2018), they were all too humble to seek gratitude in recognition or a bit of the spotlight if angled in their direction, which it never was. They'd rather retell old-boy yarns among themselves about their glory days than be crowned for their momentous achievements.

I was fascinated by their intellectual brilliance, engineering prowess, and charming humility when speaking about why they did what they did and how they did it.

I asked to speak for them. They were delighted; so, I wrote *Bright Boys*, which was published in 2010.

Because 2019 is the 70th anniversary (1949–2019) of the bright boys pinging off that first binary bit out into the world, it's a good time to stop for a moment to take note of and focus more closely on a story that's basically remained untold for a very long time.

From the first book's full title: *Bright Boys: The Making of Information Technology 1938–1958*, I have pared back the story from the 20 years it took for Information Technology to rise and take hold, in order to isolate the digital goings-on in the Barta Building, and slightly thereafter, as the action moved from Cambridge to Lincoln Laboratory in Lexington, Massachusetts, 12 miles away.

To the parings from *Bright Boys*, I have added lots of new material as well as looked a bit more closely at the original story. *The Untold Story of Everything Digital: Bright Boys, Revisited* is the end result.

Jay and Bob weren't able to help me out with my manuscript this time around: Jay passed away in 2016; Bob in 2018. I had hoped that they would have had a chance to at least read the new book. Sadly, no.

By now, I'm thinking that they trust me. So, here goes.

Chapter One

Terror at the Pentagon Taxi Stand

"Wizard War" Begins

On a scorching September afternoon at the Pentagon taxi stand in 1949, George Valley dashed to the running board of a moving taxi. The lone passenger inside, a good friend, the Caltech theoretical physicist H. P. "Bob" Robertson, had hailed him.

"George, get in," called out Robertson, swinging open the door.

Valley was hoping for a quick lift to Union Station for the train home.

He had no way of knowing that after settling back into the rear seat and exchanging a handshake with Bob that his life was about to change—suddenly and forever. He had no way of knowing that he was about to be hurtled headlong on a fantastical journey where he and a small group of engineers—engineers he was yet to meet—would make the world go digital for the very first time.

Valley was at the Pentagon attending technical meetings as a civilian member of the United States Air Force Scientific Advisory Board, known in government circles as the SAB. The 36-year-old Valley was an MIT physics professor and radar pioneer, who, during the war at MIT's famous Rad Lab (1941–1945), had led the team that developed the H2X radar bombsight.

Strategic Air Command (SAC) B-47 Stratojet bombers (c. 1951). (Courtesy of the United States Air Force.)

Valley and Robertson first met and became good friends in London during the war while on assignments together for the Vannevar Bush-led Office of Scientific Research and Development (OSRD).

A SAB member since 1946, he was frequently in Washington, as was Robertson, who was technology advisor to the Joint Chiefs of Staff and instrumental in setting up the Weapons Systems Evaluation Group for the Secretary of Defense.

During the war, Robertson was part of a scientific intelligence team that mainly focused on capturing German V-2 rocket technology before the Russians could. Fluent in German, he was famous for his team's techno expedition off the beachheads at Normandy, following directly behind advancing troops.

Together with similar units of British Commandos, with Canadian troops and Free French forces—all operating independently of one another and, most times, in competition with each other—Robertson's scientific intelligence team made a mad dash into Europe to grab up anything and everything of German technology.

One of the most celebrated of these techno booty hunters was a dashing British captain named Ian Fleming, who would later fold many of his escapades into a series of books for his secret agent hero, James Bond.

Robertson wasted little time getting to the point: "George, I've just come from a briefing about how the president intends to announce to the American people that the Soviets have detonated an atomic bomb."

Valley wasn't surprised. Everyone in the scientific community knew the day was coming fast. But this soon? "Wasn't Kurchatov supposed to be four or five years out from perfecting a bomb?"

"That's what we all thought," replied Robertson, "but Joe Stalin must have had other ideas about waiting. Truman will issue a brief statement to the media on September 23."

"Then it's been definitely confirmed?" asked Valley.

"Last month, August, seismic instrumentation pointed to a part of the Soviet Union called Kazakhstan; then airborne radiation was picked up over Siberia. Truman asked for more proof. Intelligence now confirms a detonation about 200 miles west of Semipalatinsk near the Irtysh River ... in Kazakhstan. A blast radius of 1.5 miles seems to put it in the 20-kiloton range, about the size of our Trinity test."

Valley listened and nodded as Robertson ticked off the proof. "Well, not to be flippant," he quickly interjected, "but the USSR is oceans away or at least the other side of the North Pole from our borders. When they get it airborne, then worry."

"Then it's time to worry," replied Robertson. "They've got wings."

"The Soviets have heavy bombers that I don't know about?" queried Valley.

"That's correct," said Robertson. "They flew at the Tushino Air Show. They call them Tu-4s, which are actually B-29s."

Before Valley could recover from the startling news, Robertson held up a hand to hold off any reply. "Hold on, let me explain," he said. "Your face looks like mine did at my morning briefing."

"During a bombing run over Japan, three B-29s couldn't make it back to base, so they landed at Vladivostok. The Soviets sent the crews home, but kept the planes. They then proceeded to reverse engineer every nut and bolt on the B-29, right on down to the Boeing logos in the cockpit. They even copied what had been flaws in the original manufacturing process. They overlooked nothing!"

"And it actually fits the bomb bay … and they can get it airborne?" asked Valley.

Robertson nodded.

"George," explained Robertson, "let me be plain: They can now easily fly over the Pole and drop a 20-kiliton egg right on downtown St. Louis. It probably would be a one-way mission, but hey, that's war. Remember?"

"The sky just changed," mumbled Valley, shaking his head in disbelief. "Now every time we look up, it may be cause for terror."

Years later, Valley would admit that all he could think about at that moment was his family in Belmont, very near to his office at MIT. An atomic bomb dropped on Boston would easily destroy his home and family.

"And you say your morning briefing had worse news than this awful Tu-4 revelation?" asked Valley somberly.

Robertson shrugged. "These days every briefing is just about neck and neck on the gloom meter. This morning, a question posed was how the Air Force would repel formations of these Tu-4s bearing down on the United States. There was no credible answer. It appears that there aren't enough radar stations, and the stations that do exist use antiquated equipment. And tactical communication among the stations is deplorably late ... or non-existent."

"So much for any advantage from early warning," quipped Valley.

"Of course, the president will avoid any mention of the Tu-4s; no sense making the public too crazy too soon. But George," blurted Robertson, "defenseless airspace; it's scandalous! You've got to do something about it ... and fast!"

"Me!?" replied Valley in astonishment.

"You'd be perfect, George," answered Robertson. "You're one of the world's top radar experts. Even better, when everyone else is zigging, George Valley zags. We need a zagger. If anyone can remedy the situation, it's you."

"I'm a zagger?" laughed Valley.

"Precisely!" Robertson shot back. "Remember when bomber command and the Rad Lab—when just about everyone—was betting the farm on H2S radar, you zagged. You called H2S "a bucket of crap."* You proposed X-band radar that would provide higher resolution and more detail; you put a team together and then proved yourself right with H2X radar, which then became the Rad Lab's top-priority project.

"You zagged again when redesigning the H2X antenna, and then again introducing an analog computing system for the H2X."

As Greg Goebel in his "WW2 & The Origins of Radar" says of Valley's efforts with H2X: "Valley was smart, aggressive, abrasive, and determined to build a targeting radar that worked."

"Bob," said Valley, looking out the window in surprise, "what happened to the train station? We're back at the Pentagon."

Robertson laughed at his friend's surprise. "That's because the Chief is waiting to see you."

"Vandenberg?" queried Valley. "Wants to see me?"

"Yes, General Vandenberg wants to see you immediately. You come highly recommended," smiled Robertson, grabbing the handle and opening the door. "But first check in with his vice chief, Muir Fairchild."

---

* Greg Goebel "WW2 & The Origins of Radar." https://vc.airvectors.net/ttwiz.html

Valley, a bit stunned and feeling unprepared for a meeting with the Air Force's Chief of Staff, awkwardly stepped back onto the sidewalk. Robertson closed the door after him and rolled down the window.

"Know what Winston Churchill called the race for electronic superiority?" asked Robertson. "Wizard War," he grinned, quickly answering himself. "George, you and I are wizards in this new Wizard War."

"I thought we were done with wars," said Valley.

"Me too, my good friend," he nodded with a bit of a sigh. "Me too. ... Well, if needed, you know where to find me. Good luck."

George Valley was born in New York City on September 5, 1913. At 22, he earned an undergraduate degree in physics at MIT, and in 1939 a PhD in nuclear physics from the University of Rochester, New York, which was followed by a turn as a lens designer at Bausch + Lomb's facility in Rochester.

Later in 1939, he found himself back in Cambridge as a Harvard Research Fellow, and a year later trooped in with the gang as the Rad Lab was getting underway. He stayed until the Rad Lab mustered everyone out at war's end.

In 1947, he put in a year as editor of the *Radiation Laboratory Technical Series*, which turned out to be a huge asset for postwar electronics research. And except for his ongoing SAB membership, 1949 had seemed like a good year for getting back to his family, teaching, and the good life—an ample helping of "normalcy"—after the war.

Then came Bob Robertson's, "George, get in."

One Month Later ...

George Valley was on the telephone with Washington for what must have seemed like hours. On the other end were a very interested Theodore von Karman, chairman of the SAB, and his aide Air Force Major Ted Walkowicz. It was just about the 1st of November 1949, as Valley recalled it (actually, November 8), and he had recently finished his analysis of the country's air defense preparedness. It wasn't good, and he had an earful of suggestions jumping over the telephone line.

He had been to a Continental Air Command (CONAC) radar site, which, he said "resembled one of those army camps of the Indian wars that you see in late-night movies—except Quonset huts substituted for the log cabins and jeeps took the place of horses."

The SAB had supplied him with a stack of reports on air defense, which he called "all disquieting."

He had read the proposal on a temporary "Lash Up" system that General Fairchild and his air defense expert, General Gordon Saville, had presented to Congress seven months earlier in March, which called for 75 existing, WWII-vintage radars to be literally "lashed" atop towers or platforms from coast to coast across the country.

He was aghast at the thought of such primitive radar trying to spot and track waves of incoming Tu-4s, some of which would be carrying A-bombs, while most others would be decoys. There was a need to track each and every bomber beginning hundreds of miles from any U.S. border; the Lash Up was far from adequate.

Valley told von Karman that there was "no effective radar system for low-flying aircraft. Airborne interceptor radars failed when looking down on low-flying bombers, and while ground radars worked well at long ranges against high-flying aircraft, they could not detect low flyers."

Al Donovan from the Cornell Aeronautical Laboratory worked out calculations for Valley, reporting "that a bomber flying in over the north pole region at high altitude could always detect the ground radar before the radar detected it; it could thereupon descend under the radar beam and continue undetected at low altitude."

After what he'd seen of U.S. preparedness, Bob Robertson's description of "defenseless airspace" was tragically kind. As Valley recalled later, "No one expected a hostile country to possess nuclear bombs for years to come." America felt comfortably safe nestled between its vast oceans. "Cocksure and arrogant, most of us were fooled."

The only bright light in an otherwise dismal defense plan had been his visit to the Air Force Cambridge Research Lab (AFCRL) on Albany Street just around the corner from MIT. The AFCRL, which was run by John Marchetti, who was considered the Father of American Radar, was, according to Truman's science expert, William Golden, the best military research facility in the country. Valley was impressed by the people there and their research work.

Von Karman and Walkowicz were well acquainted with Valley's observations, although not to the alarming extent that Valley's investigation had uncovered. Valley, however, just maybe had a potential solution to part of the problem, which came as a bit of fresh air into the stifling atmosphere of an America floundering at preparedness. Von Karman and Walkowicz asked Valley to prepare a report on his comments and suggestions and to send it to them.

They didn't have to wait long. Valley quickly sent them a three-page summary on the same day.

Since his backseat conversation with Bob Robertson, Valley had been quick about whatever he was doing on behalf of air defense, although he was less sure about his motives for jumping in so readily—especially when it came to anything having to do with an atomic bomb.

Valley had refused to work on the Manhattan Project, and following the war, had joined up with others "to make nuclear energy into a force for peace, not doom," he later recalled. He lobbied Congress against the May–Johnson Bill that would have placed nuclear energy entirely under the control of the Department of Defense.

Writing years later, Valley said he made "innumerable speeches to lawyer's clubs, to doctor's clubs, to chambers of commerce, to Rotary Clubs, to Lions Clubs, to the League of Women Voters, to anybody who would listen." In the end, however, his thoughts were on home and family, and the sobering realization that "the blast wave of the first bomb to hit Boston" would easily reach his almost-completed home in nearby Belmont. So, with more givings than misgivings, he sent his list off to von Karman, realizing that like World War II, he was committing himself for the duration, however long that would take.

His letter noted ten subject areas to be investigated, and stated that the investigations would best be accomplished not by the military but by a special "civilian" committee drawn up of individuals with expertise in the areas he outlined.

For speed and to facilitate the meetings, he suggested that committee members live in close proximity to one another—the New York to Boston corridor—and that the meetings take place in the Boston–Cambridge area, preferably at the AFCRL.

Twenty days later, on November 28, Valley was in Washington, DC, to attend a special meeting of the SAB Executive Committee. Muir Fairchild was there to address the group, and he read excerpts from Valley's letter.

"I was impressed by his frankness," wrote Valley later. "He asked the SAB to fix the system."

Valley's report called for large, long-range radar that sent microwaves out to the horizon looking for something to bounce off and return a signal. In addition, he noted the need for two other radar types: gap-filler radar to fill in open gaps between the long-range radar units, and height-finder radar, which determined the altitude of incoming aircraft. Between the three radars, both high-flying and low-flying, aircraft could be revealed and tracked in real time, including their latitude, longitude, speed, and bearing.

He was very aware that such continental radar coverage sufficient for North America's airspace would produce an enormous amount of real-time radar data that needed to be collected, communicated, and acted upon also in real time. Otherwise, why bother? Everything would be way too late.

He knew of no real-time calculating or computing machine or system capable of doing what was so vital to the success of air defense. It was the great unknown. Such a system had to be found or built, and quickly.

It was fortuitous that Valley's recommendations got into the hands of someone like Fairchild. Von Karman had placed Valley's letter with a doer. The next day, the SAB formalized one of Valley's suggestions by proposing that an Air Defense Technical Committee be formed to address the matter. Fairchild made sure that the proposal got quickly to Vandenberg for his review and approval.

By December 8, Fairchild had forwarded the SAB-Valley proposal for an Air Defense Committee through to Vandenberg. A letter from Fairchild to Valley on December 15 confirmed that the committee got a strong go-ahead from the Chief as well as a request to begin work "within the next few weeks." Another letter from Fairchild of the same date asked Valley to take on the chairmanship of the Air Defense Committee.

The same day, Valley wrote a four-page paper elaborating on his original letter to von Karman. "Tentative Remarks on the Task, Organization, and Program of the SAB" reiterated the ten subject areas to be investigated, and with it the SAB formed the Air Defense System Engineering Committee or, as it was commonly called, ADSEC. It would soon come to be known as the Valley Committee.

Seems Robertson had chosen well.

Even better, Valley had a bankroll to work with: Fairchild and Saville's appearance before Congress the previous March had netted them $116 million for continental air defense (adjusted for inflation, it's equal to $1.2 billion in 2019).

Valley had plenty of money—and plenty of talent, what with the combined people resources of Marchetti's AFCRL and his handpicked group of experts to serve with him on the Valley Committee.

Still outstanding on his to-do list: the great unknown.

Chapter Two
Jay's Dilemma

The Anvil of Complexity

In the spring of 1949, Jay Forrester was a very worried man, and the months ahead didn't look like they would be getting any kinder.

He shouldn't have had a worry in the world. He and his second-in-command, Bob Everett, and their bright boy engineering mates had just built the world's first real-time, electronic digital computer. It was a general-purpose, parallel machine that cranked out 10,000 calculations per second, and best of all, they had groomed it to never crash.

It should have been a time for cheers and maybe a parade. It wasn't.

Named Whirlwind, it was a massive machine that filled 2,500 square feet of an old but stately looking former commercial laundry, called the Barta Building, that sat on the corner of 211 Massachusetts Avenue in Cambridge, Massachusetts. Above it, and nearly as massive, was a cooling system without which the computer would indeed have crashed, which would have completely ruined the two years it took for Forrester and his bright boys to design and build it.

With the ubiquity of digital computing today—in a world "going digital" in every conceivable way—it's hard to imagine that Whirlwind and its creators were not more appreciated from the git-go. If the world had known then how much Whirlwind was about to affect just about everything, Forrester, in the spring of 1949, might have been fending off paparazzi rather than ill will.

Interior Barta Building and Whirlwind memory unit: Charles Corderman (on ladder); Gus O'Brien (below left); and Norm Daggett (below right). (Courtesy of MITRE Corporation.)

Sometimes even world's firsts have a tough time convincing anyone of their true value. In 1949, there was less than a megabyte of RAM on the entire planet, and every bit of it resided in the Barta Building. Forester and company had not only birthed digital computing, but they had birthed Information Technology as well. However, not many seemed to care.

There were lots of people, and powerful ones, who were totally unimpressed with what had gone on in the Barta Building. Biggest of all, Whirlwind's sponsor, the Office of Naval Research (ONR), which was in the process of paring down Forrester's fiscal budget by half … or more, and maybe even canceling its support entirely. The ONR's reason being that Whirlwind was not the flight trainer that Forrester and company had contracted to build. As such, in the eyes of the ONR and many others, Whirlwind had no defined mission, no purpose—which also caught the attention of the government's Panel on Electronic Digital Computers that was looking to eliminate government-financed computers deemed to have no mission.

There were also hard feelings from some of MIT's academics toward Whirlwind's creators. Nearly all of the discord concerned either money—that they were incorrigible spendthrifts—or assailed their credentials—that they were a bunch of no-name, junior engineers without any standing.

Somehow Forrester and Everett had latched onto a sweet contract that many felt that they did not really merit. People more senior viewed themselves to be far more deserving of the government's largesse. Kenneth Flamm in his *Creating the Computer: Government, Industry, and High Technology* ascribes the flurry of academic rancor largely to money and bruised egos, what he calls "the galling to established members of academic mathematical circles to be in close competition for research dollars with a group of young, largely unknown MIT engineers."

Of course, the facts driving the pettiness were all true. But if youth, lack of reputation, and zero previous success were removed from the equation of innovation and discovery, there would be pity few inventions in the world today.

Even deeper was the technical disconnect between MIT and the Barta Building bright boys: In 1949, when it came to digital computing, MIT's Electrical Engineering Department just didn't get it. MIT was not alone; not many others in the world "got it" either. Better to make friends than adversaries, Forrester and his band of engineering irregulars tried to educate the skeptics.

"We did have frictions," recalled Forrester years later, "trying to awaken a traditional electrical engineering department, devoted to power generation, to the arrival of computers. At one point we presented, with only partial success, a lecture trying to convince the electrical engineering faculty that it was not only possible but also desirable to use binary arithmetic for computation."

It would never get any easier. Only when Whirlwind finally cranked into action and began changing the world did minds awaken to the prowess of a digital computer. *This machine*, most everyone would later confess, changes everything.

Simultaneously trying to defend themselves and to prove themselves would take diplomacy and guerilla instincts. They were learning fast.

Even more worrisome to Forrester than disappearing financial support and disfavor from the ONR and MIT faculty was a massive technical concern looming over his project. Whirlwind's memory was stored in large, fragile glass bottles called electrostatic storage tubes that cost over a thousand dollars each to build, all by hand. With the lifespan of

Whirlwind's on-site tube manufacturing department: (left to right)—Pat Youtz, Stephen Dodd, and Jay Forrester examine the finished electrostatic storage tube, 1951. (Courtesy of MITRE Corporation.)

an electrostatic storage tube incredibly short, Whirlwind's glass bill was now eating up a large portion of the entire budget.

The 31-year-old Forrester was determined to find a solution. He had to! Whirlwind was a transformative machine, but wouldn't be for long with such an Achilles heel as fragile RAM.

On a personal level, Forrester was now a family man as well with two immediate responsibilities: a wife and daughter who also depended on his success. In July, he would be celebrating his third wedding anniversary with Susan, the love of his life who he had met when she worked in MIT's payroll department. They had a new baby girl, Judith, born in 1948, who had just celebrated her first birthday in March.

Mr. Fix It

Maybe as a consequence of growing up as a midwestern ranch boy, living far from everything with only himself to rely on to fix whatever needed fixing, Forrester had made himself into a brilliantly masterful problem solver.

Designing and building a first-ever, real-time, electronic digital computer had tried his problem-solving skills to the max, and, together with his team, he had succeeded grandly. Because of it, his bright boy compatriots revered his leadership; and maybe because of his easy going, ranch boy ways, they respected him for his generosity of spirit and kindness toward them.

Years later, bright boy Norman Taylor would say of Forrester, "It was difficult to know what he was going to do next, but he was so terribly capable, it didn't matter if you couldn't follow his reasoning. He was always thinking with seven-league boots on. It made him a pretty formidable guy to work for partly because he and Bob [Everett] always made sure you understood the problem you were working on, by finding out what you didn't know as well as what you did know."

Ridding Whirlwind of its fragile memory was Forrester's biggest challenge in perfecting his leviathan of a computer. Strangely and surprisingly, a potential solution arose one April evening in 1949, while at home idly leafing through a copy of *Electrical Engineering*.

He spotted an intriguing advertisement by the Arnold Engineering Company for a substance called Deltamax, which was a magnetic substance, a "specially treated nickel-iron alloy … developed by the Germans and used during World War II in naval fire control equipment."

Might this magnetic Deltamax somehow or other, mused Forrester, staring at the advertisement, replace the memory storage in Whirlwind's expensively unstable electrostatic storage tubes?

"The idea immediately began to dominate my thinking," he wrote later, "and for the next two evenings I went out after dinner and walked the streets in the dark thinking about it, turning over various configurations and control methods in my mind until I began to see a configuration in which we could use the magnetic element..."

That serendipitous moment over a page in *Electrical Engineering* would eventually lead him to one of the greatest discoveries in the history of computing. Forrester's discovery of magnetic-core memory would stabilize computer memory not only for Whirlwind but for any other computer across the entire fledgling industry.

Eventually, magnetic-core memory would become the industry standard until Intel's first chip in 1971. Aside from the prestige associated with the discovery, MIT netted over $25 million in royalties (those 1950s dollars tally up to over $250 million in 2019). To date, it ranks as the single largest royalty haul in MIT's history.

Close-up of magnetic-core memory. (Courtesy of MITRE Corporation.)

Of course, in the spring of 1949 magnetic-core memory was just the shimmer of an idea that he dragged through the streets on his nighttime walks. However, on Monday, June 13, 1949, he sketched out in his notebook a latticework of magnetic elements that he titled "Notes on a Magnetic Storage Method." The shimmer was gaining substance. He saw promise in that sketch, and began to pursue its realization.

But that was merely a sketch in an engineer's notebook. No matter how tantalizing it was to ponder or how doable he thought the concept, he had more pressing and weighty concerns needing to be addressed before the year was out.

He and Everett and their bright boy mates had constructed the most complex machine ever created by humans, yet it was difficult to find anyone outside of the Barta Building who thought their creation had any purpose or use.

Everything Has a Beginning
From 1939, when Jay Forrester first arrived at MIT from Nebraska until the end of World War II in September 1945, Gordon Brown's Servomechanism Laboratory was the only home that he knew. Brown

founded the lab in 1939 as a spin off from Harold Hazen's Department of Electrical Engineering.

The lab specialized in the research and development of servo-systems, which were mechanisms for positioning radar and gun-control, especially for Navy Ordnance, Army Ordnance, and the National Defense Research Committee.

Bob Everett arrived at Brown's lab after his graduation from Duke University (first in his class) in 1942.

George Valley worked nearby in the Radiation Laboratory, popularly known as the Rad Lab, which was a microwave and radar research laboratory during the war.

When the war ended in September 1945, both labs emptied out quickly. Thousands left in waves, quickly scattering to business, industry, and academia; back to civilian pursuits, peace, and the gathering postwar boom.

Forrester was prepared to leave as well. "I assumed that I would be leaving MIT," he wrote, in his foreword to the 2010 edition of *Bright Boys*, "either to find an industrial position or to start a new company to carry forward and apply our feedback system knowledge to physical engineering applications."

Brown had other ideas for Forrester. He didn't want to see the talented, 27-year-old engineer and friend skip off to parts unknown. That same September, Brown asked Forrester over to his office for a chat.

As Forrester would later explain it, Brown didn't offer him a job; rather, he handed him a piece of paper with a bunch of projects listed on it. Brown asked him to take his pick. Forrester was flattered that Brown thought so highly of him as to offer a choice of any of the ten projects listed on the paper.

"I decided to pick one from the list. Captain Luis de Florez at the Navy's Special Devices Division in Port Washington, Long Island, had proposed it."

Throughout World War II, the Special Devices Division of the Navy's Bureau of Aeronautics, created and commanded by de Florez, was a center for developing training devices for combat, like using motion pictures to train aircraft gunners or kits to build model terrains in order to facilitate operational planning in the field. De Florez, a Navy aviator and MIT grad, called them "synthetic training devices," what today are better known as simulations.

Following the establishment of the Office of Naval Research (ONR) in 1946, the Navy's Special Devices Division was renamed the Special Devices Center (SDC). The modern iteration of the SDC is the Naval Air Warfare Center Training Systems Division (NAWCTS).

The de Florez flight training device selected by Forrester was first offered to Bell Labs where it was declined before winding up on Brown's list of to-do projects. Unwittingly, the Bell Labs refusal sparked the revolution that jump-started the bright boys into digital computing. Forrester and his gifted 23-year-old sidekick, Everett, would set into motion one of the most magical adventures in the history of engineering.

"The de Florez project," recalled Forrester, "was to go beyond the existing flight trainers for pilots, which were tailored to the characteristics of a known airplane. The project was to create a cockpit that would exhibit the characteristics of a proposed plane, based on wind tunnel data of a model of the plane."

The Navy approved a preliminary design study for $75,000 for what was to be called the Aircraft Stability and Control Analyzer or ASCA. Prior to ASCA, flight simulators were built for each and every type of aircraft; this time, however, the Navy sought a single flight simulator that could be used as a general-purpose trainer for any aircraft, either existing or planned. In short, a simulator.

Beyond the preliminary contract, MIT received the Navy's go-ahead contract for $875,000, although some Navy engineers fated it to early failure, calling the project "an engineer's nightmare."

In 1945, computation meant only one thing: an analog computer, which was a colossal but intricate maze of metal gears, cams, spindles, and shafts. Quite alien looking to the modern eye, the room-sized machines looked like the giant handiwork of a clockmaker gone mad. But to contemporary engineers, they were marvelous machines of precision and eye-fetching wonder. Some were electromechanical, but even electricity could not power the gears beyond their limitations.

The ASCA system was frustratingly complex as a cumbersome analog gear works, and Forrester was unable to cope with implementing the simulator's aeronautical equations in real time.

Bob Everett with the Aircraft Stability and Control Analyzer, circa 1945.
(Courtesy of MITRE Corporation.)

Fortunately, along came Perry Crawford. Crawford was a good friend of Forrester, an MIT graduate student (Center for Analysis) who had done his 1942 master's thesis on automatic control by arithmetic means.

Perry Crawford, Forrester remembers, "was a person with continually unfolding visions of futures that others had not yet glimpsed. He was always looking, listening, and projecting new ideas into the future." Far from being the stereotypical researcher holed up in his lab, Crawford "was uninhibited, not restrained by protocol or chain of command, and a free-wheeling intervener in many circles of activity."

And as luck would have it, the "uninhibited" Crawford, the digital visionary who pushed the idea of combat information and control by electronic computers in his master's thesis, was, by October 1945, working for the Navy's Special Devices Division (soon to be the Special Devices Center [1946]) and now had oversight of the Whirlwind project.

Crawford "intervened" with Forrester regarding the potential of electronic control for his reluctant device.

To his credit, Forrester quickly agreed. The unyielding technology had to be ditched, a new plan for success devised for the customer, and momentum maintained for MIT management as well as the project's 15 full-time research engineers and 50-odd graduate students who looked to Forrester for leadership.

If Forrester ever pined for a smattering of digital reality, he need look no further than his attendance at Raymond Archibald's *Conference on Advanced Computation Techniques* held at MIT the last three days of October in 1945 where, among other things discussed, was that of the great ENIAC computer, purring away at the Moore School of Engineering at the University of Pennsylvania.

Forrester had heard enough; he had to see what ENIAC looked like up close and personal, so he made a quick trip to the Moore School to kick the tires on the giant machine.

And giant the massive machine most surely was. It was 8 feet tall and 80 feet long. It had 70,000 resistors, 10,000 capacitors, and a half-million soldered joints. Its 30 tons took up 3,000 cubic feet of space, and when its 17,468 vacuum tubes powered up, the room temperature jacked up to 120°F. Rumor had it that lights in sections of Philadelphia would flicker.

The result of a $400,000 contract with the Army in 1943, the computer's purpose was to calculate gun-firing tables for the Army's nearby Aberdeen Proving Grounds. The trajectory of an artillery shell like the 155-millimeter "Long Tom" covered over 500 sets of conditions. A team of women punching numbers into mechanical calculators would take more than a month to calculate a single table—a table without which the gun could not be properly aimed and fired.

ENIAC was a decimal machine. It wasn't real time or general purpose; it calculated in serial and was programmed by plugging in wires by hand for each calculation. But the hulking computer was indeed fast, over 1,000 times faster than any mechanical device. A trajectory that formerly took 20 hours to tabulate on a desk calculator, ENIAC could kick out in 30 seconds.

Forrester was staring at the future. It was exactly what the ASCA project needed.

In late August 1946, the Army would dismantle ENIAC and cart it off to Aberdeen and its Ballistics Research Laboratory where it performed faithfully until final shut down in October 1955. Forrester had seen it just in time.

While hobnobbing at the Moore School, Forrester discovered upcoming plans for an even newer electronic machine called EDVAC (Electronic Discrete Variable Automatic Computer). As a consultant on the ENIAC project, the renowned mathematician John von Neumann had come up with a document called "First Draft of a Report on EDVAC." EDVAC's machine design was superior to ENIAC's in that EDVAC introduced a revolutionary new idea, that of the stored program concept.

"Stored program" means that the computer's memory possesses not only data, but also the instructions that the computer uses to manipulate that data. ENIAC's instructions were programmed by plugging wires into the machine and then having to rearrange the wires with each change in programming. A simple program change could take hours.

John von Neumann revolutionized that in a hurry. His report would become a seminal document in the history of digital computing. Every computer today is a stored program computer, and the innards of every computer are known as von Neumann architectures.

It would become another essential addition to the ASCA project. Forrester and Everett had in mind to build a machine bigger, better, and faster than either ENIAC or EDVAC.

The first step in the conversion process would be the risky maneuver of switching the Navy from analog to digital electronics. The Navy might well have balked, finally agreeing with its own engineers who had earlier cursed the project as impossible. Presenting the Navy with a good alternative plan, a little brinksmanship with some able help from Crawford, and together they prevailed.

Forrester's new proposal to the Navy called for the building of an electronic digital computer and a flight simulator/analyzer for $2.4 million to be delivered by 1950. The Navy accepted the proposal, funding the construction for an initial $1.2 million through mid-1948.

It was quite the sales job by the young engineer; Raytheon, a company that Vannevar Bush helped to found, offered to build a computer for the Navy for the bargain price of $650,000. Forrester persuaded the Navy to fund his electronic offering, saying that "the potential of the digital computer was so great and the benefits derived from its use so immense that the costs involved, no matter how great, were warranted."

The Navy bought into the plan, naming it Project Whirlwind. The name Whirlwind was selected following the Navy tradition of naming its computers after air movements: Hurricane, Typhoon, and Zephyr were other Navy computers.

Forrester and his bright boys were set loose to construct their real-time, general-purpose electronic digital computer. It would be the world's first.

Forrester and Everett gathered their crew of young, bright boy engineers together. Forrester told them to put an immediate stop to their work. From here on out, he announced, we are going to design and build a real-time, electronic, digital computer.

As Forrester had expected, a sudden wave of puzzlement uneasily flitted over their faces.

He explained that the new machine was to be designed and ready to build by the end of 1947, with the buildout process running the whole of 1948. Thereafter, he continued, they'd knock any kinks out and have it up and running by the end of 1949. Any longer than that, he thought, and the Navy would grow impatient as would the management at MIT.

Finally, in addressing their questioning looks, he told them that no one had ever built the kind of machine that they will soon undertake. Not even the builders of ENIAC.

With a bit of a wry smile and a confident nod toward the group, the midwestern Mr. Fix It in him finalized: Guess that makes all of us pioneers. Now, let's get busy. Let's not be late for the frontier.

First item on their agenda was to learn a thing or two about electronic digital computers—and to learn really fast!

Chapter Three
The Whistle Factory

Jay and Bob

In 1946, barely two dozen people in all of the United States knew what an electronic digital computer was all about, and none of the bright boys were among those elite few.

For Forrester and Everett, the great ENIAC at the University of Pennsylvania's Moore School of Engineering was the epicenter of electronic computing and quite possibly held the key to Whirlwind's future success. Their invitation to a special course offering on electronic computing at the Moore School seemed like a perfect way to kick off their new venture with Whirlwind.

Jay Wright Forrester was born in 1918 into the great Midwest, specifically that northwest chunk of Nebraska known as the Sand Hills, a huge, swirled expanse of ancient sand dunes topped over with a thin crust of topsoil and grass. Grass, grass, and more grass for as far as the eye could see across Custer County; hundreds of square miles of it filling everything between the North Platte and Niobrara Rivers, 20 miles from the nearest town of Anselmo. Just right for cattle ranching.

An ocean of rolling grasslands also just right for the sojourns of a young boy's imagination. Unlike his rancher father, Duke, Jay had a decided mechanical bent and put it to good use on all manner of helpful contraptions, like in his senior year of high school, a wind-driven system that brought the first-ever electricity to the farmhouse; light bulbs, radio, even a toaster

(Detail) Moore School of Electrical Engineering, University of Pennsylvania, c. 1947. (Courtesy of Collections of the University of Pennsylvania Archives.)

were then possible … as well as an electrified screen door to zap bugs. It may not have been the Tennessee Valley Authority, but to a youthful sense of accomplishment, it was mighty close.

Forrester was a tall, gangly kid of those Nebraska grasslands, the same endless prairie that Willa Cather wrote about in *O, Pioneers!* where "Under the long shaggy ridges, she felt the future stirring."

With a penchant for tinkering with all sorts of machines, devices, and mechanical ideas, ranching on the grasslands was not stirring any sort of future for young Forrester. College took him clear across the state to the University of Nebraska at Lincoln. Tuition was $35 a term, and 1939 saw him graduate with a degree in electrical engineering. With his newly minted degree in hand, the 20 year old lit out for the halls of MIT, a place from which he would never leave. Goodbye, Nebraska.

Nearly 1,500 miles from the Nebraska Sand Hills but just 10 miles from the neatly trimmed outfield grass in Yankee Stadium is the decidedly East Coast melting pot of humanity on the Hudson River—Yonkers, New York, the birthplace of Robert Rivers Everett in 1921.

For another lad with a mechanical bent, Yonkers had the perfect atmosphere for inspiration into the engineering side of things. The world-famous "electrical wizard," Charles Proteus Steinmetz, electrified the Yonkers' trolleys; Edwin H. Armstrong invented FM radio broadcasting there; Leo Baekeland chose Yonkers to cook up his first batch of synthetic plastic, which he named after himself, Bakelite; while the first practical dynamo and first practical electric engine were both invented and produced in Yonkers.

Then too, Everett's father was an engineer, a civil engineer, which the young Everett desperately wanted to avoid becoming. Electrical engineering was going to be the calling for this son of Yonkers. Easygoing, affable but quiet, and very bright, young Everett journeyed to Duke University, graduating first in his class in electrical engineering in 1942, immediately following which he set himself down at MIT for more of the same.

Both Forrester and Everett were fortunate to find themselves in 1942 as members of Gordon Brown's Servomechanisms Laboratory, which was a next-door neighbor to MIT's microwave wonder factory, the Rad Lab.

Forrester had helped Brown to found the Servomechanisms Lab in 1940, becoming co-director the same year. (The neighboring Rad Lab began operations in June 1940.) Everett was a graduate student for a semester before heading off for a permanent job at GE. He never got to GE; Forrester hired him into the Servomechanisms Lab.

Servomechanisms were high priority during the war, especially when combined with tracking radar and anti-aircraft guns.

Without a servomechanism it took 14 people to detect an aircraft, calculate its future position, and then to precisely move a gun, aim, fire, and hopefully hit the moving aircraft. Against fast-moving fighter planes, such a manual system was woefully inadequate. The famous SCR-584, developed by a Rad Lab team led by Ivan Getting and Louis Ridenour, successfully combined tracking radar, a mechanical computer, and servomechanisms for a deadly accurate anti-aircraft system. Used against 104 V-1 rockets fired by the Germans at London, the SCR-584 shot down 100.

The close proximity of the Rad Lab filled with physicists working to develop microwave radar also helped to broaden Brown and his young engineers' understanding of pulse circuitry.

Getting, a physicist who claimed to be the second or third employee hired at the Rad Lab, said that 90 percent of the 2,000 professional staff there were also physicists. The nature of working with microwaves was the reason for their great numbers because "a physicist's education in the fundamentals of electromagnetic theory and associated subjects," said Getting, "was more widely embedded than it was in the professional electrical engineer of the time":

> Professional electrical engineers generally spent time working on sixty-cycle power equipment or power transmission or transformers, or worked in radio waves modulated at sound frequencies, say up to ten kilocycles. They were not versed in microwaves at all.

Such a handy dose of experience would go to benefit Forrester and Everett's later design efforts with Whirlwind. Everett, who worked on radar antenna mounts at the Servomechanisms Lab, later recalled that "radar technology was coming out of World War II activities in tubes and pulse circuits and storage devices that had been developed for moving target indicators. The technological foundation of pulse circuits made possible electronic digital computers."

Projects like the SCR-584 also brought into sharp focus the need for "teams" to complete complex machines: interdisciplinary teams of physicists, mechanical engineers, and electronic and electrical engineers were necessary to pool their talents and expertise.

Getting's SCR-584 automatic tracking radar required servomechanisms that did not yet exist. It required engineering of very skilled, very specialized mechanical antenna mounts, which Forrester and Everett built while working with the Rad Lab's physicists. And they learned as well the ins and outs of high-frequency pulse circuits "more in the television type of one to ten megacycles," said Getting. "You would find it difficult to run into more 'run of the mill' electrical engineers, who had designed circuits and abused vacuum tubes way beyond their specifications."

This on-the-job teamwork would bode well for Forrester and Everett when much later they needed to coax optimum performances from Whirlwind's thousands of chancy vacuum tubes. Team members also directly shared their knowledge at group sessions designed to spread ever-newer technologies as soon as they sprouted in either or both labs.

This team approach and cross-fertilization of ideas was not lost on the young Forrester and Everett. Although the Rad Lab closed for business in 1945 and its staff quickly went their individual ways, its early-on dabbling in a systems approach to complex projects would be fully embraced and refined by Forrester, Everett, and their band of bright boys.

Forrester and Everett transferred this wartime team concept directly into the Whirlwind development process. Recalled Forrester later, "We had a core team of engineers who had been through several cycles from basic research and on into development, design, production, and solving equipment failures in the military field. They knew each other's strengths and weaknesses and they understood the entire process of bringing an idea to field operation."

Future results bordered on the fantastic: their electronic digital computer would eventually average an unprecedented 99.8 percent uptime—and do so for a continuous 28 years. Years! Today, with computer models changing as rapidly as every six months and with computer crashes a frequent nightmare, the quality and longevity of Whirlwind-developed technology is still without equal. And the systems approach that they methodically refined soon became the crown jewels of engineering, and in some manner or form was purloined by every military or commercial organization with a complex project on its hands.

"I doubt it would be possible today," commented Forrester on his team's success, "to design a system of that complexity in so short a time and at so little cost."

But all of that was still to come for Forrester and Everett. In the summer of 1946, it was just two young guys in the full headed down the road to the Moore School and their destiny.

The Day the Gears Stopped

Hastily organized two months after ENIAC's coming out party (top secret until February 14, 1946, *New York Times*) and financed by Army and Navy Ordnance, the Moore School lectures were by invitation only to 28 individuals.

In ENIAC, the military had witnessed firsthand the promise of electronic digital computing and now wanted to push forward the development edge with the hope of getting more of the same or better. Maybe by financing lectures to a handpicked audience, a few new gems might soon glitter into existence.

Moore School of Electrical Engineering, University of Pennsylvania, circa 1947. (Courtesy of Collections of the University of Pennsylvania Archives.)

ENIAC's reputation was well known in the Pentagon, and the Moore School lectures were high on its to-do list. Two top-secret military intelligence groups sent representatives to the Whistle Factory. Having already shown its mettle in gun-firing calculations and atomic research, ENIAC—or an ENIAC-like machine—might well do the same for information gathering and analysis.

Military code breakers from the OP-20-G (Office of the Chief of Naval Operations, 20th Division of the Office of Naval Communications, G Section) and the Army Security Agency were in attendance. The OP-20-G's James Pendergrass returned as an electronic convert and vigorously pushed his Moore School lessons on the Navy. Both the Army and Navy were well aware that if the postwar military presence abroad was reduced to a police force, then an army of eyes and ears was necessary to gather intelligence in its place.

With the Navy's blessing for Whirlwind in hand as well as a fatter contract, and with MIT's concurrence, Gordon Brown's youthful *digerati* headed south to Philadelphia in the summer of 1946.

It would turn out to be an eventful journey filled with discovery and enlightenment. Held from July 8 to August 31, the course, *The Theory and Techniques for the Design of Electronic Digital Computers*, was both attended by and taught by a veritable Who's Who of the founding fathers of modern digital computing, all of whom, except for John von Neumann, were total unknowns at the time. For Forrester and Everett, two young guys struggling to get their electronic creation off the ground, this was an important intellectual bazaar filled with new ideas, insights, and techniques all handed up by the originators of an onrushing digital future.

The lectures were the death knell for analog computing. The Moore School's 48 lectures on going digital with electronic computers were telling for their absence of anything relating to gears. And everyone attending the Moore School lectures knew it and had come to witness, among other things, the first jarring screech in the slowdown.

It is doubtful that without ENIAC the Moore School lectures would have happened at all. ENIAC had charisma, and its success had world-class pull. Although monstrous and flawed, ENIAC was real. It was not a rack of electron tubes hidden away in a research lab or a half-baked prototype gone half wrong or even a theoretical paper with an eye to the future.

ENIAC was as big as life; you could pet it. And it worked. In a very real sense, it was like any other successful commercial product: It was bought and paid for by a customer, worked to the customer's delight, and now the customer wanted more.

Thomas Edison's movie camera was a beast of a thing, but it was a first and it worked. The movie camera forever changed how people acquire and record information. In the long run that is all that history really remembers: events and things that reshape the world. ENIAC was just such a change agent. And everything that preceded it, no matter how inventive, innovative, or a stretch of genius on the road to electronic computing, paled in its shadow.

ENIAC pushed into being one of those signal events when technological change is so startlingly new that, as Marshall McLuhan put it, the entire *sensorium* changes: The brain needs to turn and realign itself to properly comprehend what's happening. *Sensoria* everywhere would take some time to reorient to the full drift of digital electronics.

This was the dawn of electronic digital computing—its Iron Age of huge, hulking cabinets crammed high with components, thousands of hot electron tubes, and miles of hand-wired connections. And ENIAC was its first lord of the iron. Without ENIAC and its success, the Moore School lectures would have been hard pressed to be anything else but more visionary ramblings on what the coming electronic future of computing might portend. ENIAC was electronic computing's Big Bang; an impact heralded as big stuff on no less a public stage than the front page of the *New York Times*.

ENIAC commanded attention, and its prowess commanded attendance at the lectures. Everyone came. A team of Moore School staff led by two no-names, physicist John Mauchly and his youthful engineering pal J. Presper Eckert, had painstakingly built ENIAC into what computer pioneer and historian Herman Goldstine would later call the most complex machine ever built. It was now hoped that 28 handpicked Moore School invitees would take up the challenge to way go beyond ENIAC. Forrester and Everett were but two—but a very important two. Whirlwind would in short order change everything that preceded it and hurry the integration of computing machines and people into everyday life. With Whirlwind, computers became tools, and not just calculating phenoms.

The grinding to a halt of analog computers was inexorably complete over a remarkably short period of time. Such a brief lifespan gives pause for thought about their true identity. Were they really computers or, as the British computer scientist, Maurice Wilkes, suggests, merely calculators—calculators grown gargantuan as they approached the limits of their capabilities?

ENIAC's first big postwar job was a calculation in nuclear physics that would have taken 100 human years to solve by conventional methods. ENIAC solved the problem in two weeks (two hours of which was for the calculation, the balance spent in checking the results), and ENIAC was highly accurate. Any answer going out lots of decimal places was risky business for an analog computer, but for an electronic digital machine it was a piece of cake. And therein resided most of the reason for its demise: Speed of calculation and increased accuracy every time out was the postwar order of the day. And faster still was more like it.

By today's standards, ENIAC's two hours or even two minutes were way too long to assist something like jumbo jets stacked up by the dozens over Chicago's O'Hare Airport and landing at the rate of one every two minutes, or for banks needing to accurately process billions of checks daily. Each demands speed and accuracy—and to do both in real time.

Whirlwind would be all three at once: fast, accurate, and real time. And that was the electronic siren call of the Moore School that sent the youthful Forrester and Everett winging south in the summer of 1946.

The summer of 1946 witnessed a changing America as 11 million soldiers, sailors, and airmen tried to reacclimate themselves to their homeland after a brutal war that had seemed like it would never end. Sadly, nearly 400,000 young Americans would never return to enjoy the summer of 1946 or any other season. Forrester and Everett were veterans of the desperate effort to shorten the war from their laboratories, building tools for war that they hoped would hurry the end along.

Wartime rationing was over, and 15 million radios were produced, keeping all the returnees informed about Vaughn Monroe's latest on *The Hit Parade* or about a whole lot of babies that were being born in what was to be the first year of the Baby Boom. There was even a manual produced on how to raise all these kids: Dr. Benjamin Spock and Pocket Books put out the 25-cent bible on child rearing, *The Common Sense Book of Baby & Child Care*, which eventually sold 50 million copies in 39 languages.

Radio was about to lock horns with the new media upstart, television. Families crowded around miniscule 5-inch black and white screens for

Ted Mack's *Amateur Hour*; or the first-ever TV soap opera, *Faraway Hill*, on the DuMont Network; or sports like the Joe Louis–Billy Conn heavyweight bout beamed out to 100,000 on *The Gillette Cavalcade of Sports*.

Jobs killing the enemy for just over $70 a month had been plentiful, but for returning GIs jobs were mighty scarce at home. Those who had work averaged $2,500 annually (about $208 a month—some $138 more a month than for their battlefield heroics), and for half that salary anyone could pick up a snappy Plymouth sedan and fill the tank for 15 cents a gallon. In New York City, the Beat Generation kicked off when Jack Kerouac and Allen Ginsberg first encountered the charismatic Neal Cassady, who inspired both of their influential literary classics, *On the Road* and *Howl*.

Although the first 30,000 home air conditioners rolled out of America's factories to beat that summer's heat wave, an air-conditioned movie theatre was the best bet for a few hours respite. Millions of Americans did just that. In record numbers not seen since *Gone with the Wind*, moviegoers jammed in to catch the year's cinema blockbuster *The Best Years of Our Lives*. The Samuel Goldwyn classic about three returning servicemen in the aftermath of World War II, based on a *Time* magazine pictorial from 1944, explored the troubling irony that for many servicemen "the best years of our lives" were spent overseas in wartime and not upon their return home.

The starkly realistic and poignant film won the best of nearly everything at the Academy Awards, grabbing seven Oscars. Stiff competition at the 1946 Oscars came from other soon-to-be classic films, *The Razor's Edge*, *The Yearling*, and *It's a Wonderful Life*, which all seemed to explore the understandable postwar themes of soul searching, introspection, self-sacrifice, and right doing.

Ferried to their hometown in the bombardier's space of a B-17, the three veterans in *The Best Years of Our Lives* look down on a graveyard for bombers. Fred, a former B-17 bombardier (played by Dana Andrews), is stunned: "…they're junking them…Boy, oh boy, what we could have done with those in '43! Some of 'em look brand new, factory to scrap heap. That's all they're good for now."

Soon to get chucked upon the very same postwar scrap heap would be analog computing, and the Moore School lectures would have a major hand in lifting it to the very top of the pile. It was out with the old and in with the new, and the bright boys' plan for Whirlwind would help to usher in a new age of digital technology.

The careers of analog engineers by the thousands, those that could not or would not accept the new world order and retrain themselves for it, met untimely deaths along with their machines. "The rise and fall of this profession," sadly recalled George Valley, "is a poignant story—of expectations that came true for other people, of ruined careers, of competent engineers pushed down to technician level."

By contrast, the Moore School's electronic digital world was unseen and untouchable—everything happened deep within a box. The discovery of the transistor was yet another two years off, still further out were integrated circuits and modern computer memory. Back then, memory for data, and later, for commands, was held in strange places like mercury delay lines or magnetic drums or electron tubes.

Getting commands into ENIAC—programming—was accomplished by arranging a set of wires to one end of the machine. Switching on the electricity—a lot of electricity—would then provoke magic to happen inside ENIAC's box of tubes and circuits, which in turn pushed an answer out the other end. All done without gears. Strange stuff for the comfort zone of an analog engineer to abide.

Every computer today must be a real-time machine, because no military commander, no air traffic controller, no bank, no insurance company, no merchant of any kind, not even Junior doing his homework, waits for answers.

In 1946, no one wanted or expected a computer to spit out answers in real time; they were thankful for a two-hour wait with ENIAC, which was far preferable to holding hands with a calculator for 100 years. ENIAC was plenty fast for them and cracking tough equations was enough of a result.

Computers doing real jobs like tracking waves of enemy bombers or automating a factory or trading a billion shares a day at the stock exchange just did not compute for most. Forrester and Everett, who had spent a war building complex yet practical machines for desperate times, knew better.

Cathedral and Bazaar

Around campus, the Moore School of Engineering at 33rd Street and Walnut was known as the Whistle Factory. Some say the name came about because of the noises emanating from ENIAC, others because it was the former home of the Pepper Musical Instrument Factory from 1911 to 1926.

During the summer of 1946, the Whistle Factory was the site of the first-ever conference on electronic digital computing. It was the late forties, when there were very few conferences in science and technology in general, and fewer still in computing. And none in electronic digital computing! A summer at the Whistle Factory was a rare and remarkable occasion for the few existing *digerati* to come together to pool their talent and expertise.

Most everyone was from the East and industry or university affiliated, some trickled in from California, and still others arrived from England. Getting anywhere in 1946 was arduous; one could not just jump a quick jet for Philadelphia. The closest thing to air travel was leapfrogging in a twin-prop DC-3. The U.S. interstate highway system was still more than a decade off, but gas was cheap and the scenery not too bad for the drive south. That left buses and trains for long distances, or for hopping oceans, the lumbering DC-4.

Still, 28 attendees came for the 48 lectures. Each lecture was given in the morning at the rate of five per week, each lasting about three hours, which was followed every afternoon with an informal discussion on that morning's topic.

"It was an opportunity to tell the world what we'd been doing," recalled Charles Covalt Chambers, a Moore School professor of engineering (dean in 1949), "and it gave a lot of people training."

Electronic computing had no learned journals in which to publish. Herbert Grosch, who was the leader of IBM's Watson Laboratory in 1945, said that "what professors call 'the literature' there wasn't any—at least for computers…the exception, MTAC (Mathematical Tables and Other Aids to Computation), published quarterly at Brown University. In the Thirties and Forties computer articles were turned away by engineering and scientific journals unless well sponsored, even MTAC was careful."

The ENIAC team came together in full force for the lectures. J. Presper Eckert delivered eleven of the lectures; John Mauchly six; Herman Goldstine six; Bradford Sheppard six; Arthur Burks three; Kite Sharpless two; and Hans Rademacher and Jeffrey Chu one each. John von Neumann contributed a lecture, as did Irven Travis (supervisor of research at the Moore School [1946–1948]).

An impressive lineup of nine other future digital notables also contributed as lecturers. Forrester's good buddy, the irrepressible Perry Crawford, was one. The complete list of lecturers and their topics displays the comprehensive approach taken by the organizers at the Moore School "to tell the world what we'd been doing" and to "give a lot of people training" in the dark arts of electronic digital computing.

George Stibitz Bell Labs	Introduction to the Course on Electronic Digital Computers
Irven Travis Moore School	The History of Computing Devices
John Mauchly Electronic Control Co.	Digital and Analog Computing Machines
Derrick Lehmer Univ. Calif., Berkeley	Computing Machines for Pure Mathematics
Douglas Hartree Univ. Manchester (UK)	Some General Considerations in the Solutions of Problems in Applied Mathematics
Herman Goldstine Inst. of Advanced Study	Numerical Mathematical Methods I
Herman Goldstine Inst. of Advanced Study	Numerical Mathematical Methods II
Arthur Burks Inst. of Advanced Study	Digital Machine Functions
John Mauchly Electronic Control Co.	The Use of Function Tables with Computing Machines
J. Presper Eckert Electronic Control Co.	A Preview of a Digital Computing Machine
C. Bradford Sheppard Moore School	Elements of a Complete Computing System
Herman Goldstine Inst. of Advanced Study	Numerical Mathematical Methods III
Howard Aiken Harvard University	The Automatic Sequence Controlled Calculator
Howard Aiken Harvard University	Electro-Mechanical Tables of the Elementary Functions
J. Presper Eckert Electronic Control Co.	Types of Circuit: General

Kite Sharpless Moore School	Switching and Coupling Circuits
Arthur Burks Inst. of Advanced Study	Numerical Mathematical Methods IV
J. Presper Eckert Electronic Control Co.	Continuous Variable Input and Output Devices
Sam B. Williams Bell Labs	Reliability and Checking in Digital Computing Systems
J. Presper Eckert Electronic Control Co.	Reliability and Checking
C. Bradford Sheppard Moore School	Code and Control I
J. Presper Eckert Electronic Control Co.	Code and Control II: Machine Design and Instruction Codes
C. Bradford Sheppard Moore School	Code and Control III
Calvin Mooers Naval Ordnance Laboratory	Code and Control IV: Examples of a Three-Address Code and the Use of "Stop Order Tags"
John Von Neumann Inst. of Advanced Study	New Problems and Approaches
J. Presper Eckert Electronic Control Co.	Electrical Delay Lines
J. Presper Eckert Electronic Control Co.	A Parallel-Type EDVAC
Jan Rajchman RCA	The Selection
Calvin Mooers Naval Ordnance Laboratory	Discussion of Ideas for the Naval Ordnance Laboratory Computing Machine

J. Presper Eckert Electronic Control Co.	A Parallel Channel Computing Machine
C. Bradford Sheppard Moore School	A Four-Channel Coded-Decimal Electrostatic Machine
Kite Sharpless Moore School	Description of Serial Acoustic Binary EDVAC
John Mauchly Electronic Control Co.	Accumulation of Errors in Numerical Methods

Lecture list courtesy of John R. Harris, Virtual Travelog, San Francisco, CA.

There was an attempt to memorialize all 48 lectures for later use. It was a grand thought, since so little publishing or conference support was accorded electronic digital computing in the early days of the technology. A published record of the lectures might well be useful to many.

Each three-hour morning lecture was recorded using a wire recorder, which was a chunky, box-like machine atop which was inserted a 3-inch diameter by 1-inch spool of steel wire, much like fishing line. An ancestor of oxide-based magnetic tape, the thin wire was dragged over a recording head while the lecturer spoke. Each spool with 7,200 feet of wire at a recording speed of 24 ips resulted in one hour of audio recording.

The plan was for each lecturer to go away with his wire recordings and notes and produce a manuscript for publication. It would have made for a nifty boxed set yet didn't go as planned. Some of the lecturers followed through, others took what seemed like forever to produce a manuscript, and still others did nothing. Between late 1947 and June 1948, four volumes were published. Some volumes did not fair too well, like Volume IV, which was and still is missing nearly half of its lectures.

The great takeaway for the lecturers and attendees alike was what each brought to the others—a morale-boosting, intellectual camaraderie oozing with the firm conviction that they were involved in the creation of something truly magnificent. Within the drab confines of the Whistle Factory, electronic digital computing was alive and well; a working ENIAC was there in all its glory, while plans and ideas were afoot everywhere for newer and better machines.

The Moore School lectures lifted electronic digital computing out of its isolation. While its collective intelligence poured out, its collective eye watched, evaluated, contributed, and grew.

It was Eric Raymond's "cathedral" builders hooking up with "bazaar" builders to assess, refine, and fix their mutual electronic calling. Raymond's famous 1998 essay, "The Cathedral and the Bazaar," on the building and debugging of software, puts software development into two distinct camps. There are the cathedral builders, who are a tight-knit, dedicated few working through all of the software's problems and then releasing it to the world— the standard model for software development—analogous to ENIAC's core development crew. And then there is the bazaar-building model where software leaps out into the world and actively seeks after thousands of codevelopers—working online—to scrutinize, rework, and to suggest corrections. This was Linus Torvald's method for LINUX development— similar to the dozens of eager intellects gathered in the Whistle Factory probing each lecture and lecturer.

Raymond's view is that "given enough eyeballs, every problem will be transparent to somebody." In microcosm, the Moore School was both a cathedral and a bazaar for early hardware and programming that resulted in a much-needed outflow of new advances for an emerging technology.

And there were eyeballs aplenty in the Whistle Factory. Maurice Wilkes, over to attend from Cambridge University, cast a glance at the EDVAC's stored-program concept, then returned to England and built the first operational stored-program electronic computer, his EDSAC or Electronic Delay Storage Automatic Computer.

Sam Alexander, chief of the Electronic Computer Laboratory for the National Bureau of Standards, was there with four of his crew, notably Edward Cannon, who would produce the U.S. government agency's first electronic computer, the SEAC or Standards Eastern Automatic Computer.

In August, an ENIAC part-timer, the American mathematics professor Harry Huskey, would leave for England and Manchester University to work with Max Newman and England's resident computer genius, Alan Turing, on another of electronic computing's firsts, the Automatic Computing Engine or ACE. A year later, Huskey was at UCLA's Institute for Numerical Analysis pioneering SWAC or Standards Western Automatic Computer, the newer brother of the National Bureau of Standard's SEAC.

Kite Sharpless got command of the EDVAC project. Von Neumann and company churned toward completion of the Institute for Advanced Study's electronic machine. David Rees, a veteran of the top-secret wartime Colossus project at Bletchley Park, and I. J. Good, a Moore School lecturer, returned to England to sketch out and build the electronic Small-Scale Experimental Machine or SSEM, known at Manchester University as the "Baby."

Mauchly and Eckert's very next offering was BINAC—or Binary Automatic Computer—for the Northrop Aircraft Company. There from the atomic energy facility in Oak Ridge, Tennessee, was Cuthbert Hurd, soon-to-be leader of IBM's Applied Science Department, from which he pushed the reluctant calculator company into the world of electronic computing.

Nestled in among the attendees was 30-year-old Claude Shannon, a Bell Lab electrical engineer and mathematician, who as an MIT graduate student had worked with Vannevar Bush and his Differential Analyzer. At the lectures, he was already famous for his MIT master's thesis, which he wrote in 1937. Entitled "A Symbolic Analysis of Relay and Switching Circuits," it was central to all of electronic computing by pointing out the identity between the symbolic logic and the binary values 1 and 0 of electronic circuits.

By 1948, Shannon would become the founding father of information theory and the age of electronic communications. His paper, "The Mathematical Theory of Communication," based on his discovery of the similarity between Boolean algebra and telephone switching circuits, became the basic theory that explained the communication of information.

By August 31, *The Theory and Techniques for the Design of Electronic Digital Computers* was over, and its lecturers and attendees still had a lot of convincing to do to get everyone over to their way of thinking.

Maurice Wilkes saw Eckert as the lightning rod of the lecture series. "People who change the world," he said of Eckert, "stand facing two ways in history. Half their work is seen as the culmination of efforts of the past; the other half sets a new direction for the future."

Much the same can be said of the lectures themselves; a new direction was indelibly stamped on everyone there. They came as individuals and were galvanized into an electronic brotherhood. Everyone went home, quickly built machines, quickly spread the word, and quickly pushed electronic digital computing into the light of day. Issuing forth from laboratories both in the United States and England, no fewer than five new electronic, digital computers, the sons of ENIAC, powered up for all the world to see.

Mauchly and Eckert would produce the first commercially available number-crunching wonder machine, UNIVAC, which would make a guest appearance on Edward R. Murrow's TV show, go on to predict Eisenhower's victory over Adlai Stevenson, and quickly become the synonym for computer in every household in America.

Forrester and Everett understood their part in it all; they had to return to Cambridge to design and manage the coming out of their electronic digital machine and to cut it in the direction of real-time, general-purpose computing. Their plan was to make Whirlwind into a tool for engineers with myriad uses and potential, and not just a number-cruncher for mathematicians.

Beyond that there was not much of a plan, but that was more than enough of a beginning.

Chapter Four
The Buildout

Back from Philadelphia

Forrester and Everett returned from the Whistle Factory to Brown's Servomechanisms Laboratory raring to get going at engineering their electronic digital computer. The balance of 1946 and all of 1947, they projected to be all about designing their machine. The buildout phase was slated for 1948.

The day-to-day leadership fell to Forrester and Everett, both young (in 1947, 29 and 26 years old, respectively) and untested on large management projects. Forrester would be the battle shield and mouthpiece for the group.

As Kent Redmond and Thomas Smith noted in their *Project Whirlwind: A Case History in Contemporary Technology*, his style worked, but it drew harsh criticism and from some, even personal animosity. "Dedicated to Project Whirlwind and determined to secure its success, Forrester aggressively and single-mindedly pursued the course which he believed would most quickly reach that end."

Rather than arrogant and high-hat, he was strong willed and convinced "that the social structure and management of an enterprise are far more important than the underlying science. If one has the right environment, it will produce the required science. But the best science will fail in an unfavorable social and managerial setting."

As bright boy Norman Taylor said of them years later: "Bob Everett was relaxed, friendly, understanding and I have never seen anyone who could go to the heart of a problem so fast! Jay was as fast, maybe faster, but he was always more

Bob Everett with the Aircraft Stability and Control Analyzer (c. 1945). (Courtesy of MITRE Corporation.)

formal, more remote somehow, and you weren't always sure how dumb he thought you were, or how smart."

From the project's outset, Forrester's innovation and deliberate organizational moves seemed shrewd and professional. He set up his laboratory for ten divisions: seven for technical research and three for support.

With the help of Nat Sage, the well-respected and experienced director of MIT's Division of Industrial Research, he sought to set up an extraordinary staff. "The general type of man whom we need," he explained to Sage, who was to cast about for warm bodies to fill each slot, "should have originality and what is often referred to as genius. He should not be bound by the traditional approach."

In addition, as Forrester explained it much later: "Our relationship to the Electrical Engineering Department gave me access to incoming graduate student applicants to become full-time research assistants. They were allowed to take two subjects per term toward their degree. By reviewing applications myself, I effectively had first choice of those applying to MIT."

By 1947, the ranks of the Whirlwind bright boys had swelled. Forrester and Everett recruited 50 staff, half of whom were experienced engineers, and all, hopefully, were the kinds of people they sought: those unbounded by "the traditional approach."

"We accumulated an impressive group," wrote Forrester. "Many graduate applicants were coming back from military service with a maturity and real-world experience that is unusual for students. They were ideal recruits for the coming information age. Their MSc theses on aspects of computers qualified many to present papers at the national and international technical conventions."

In his foreword to *Bright Boys*, Forrester reflected back through the decades on his Whirlwind team and its formation. "We came out of the Servomechanisms Laboratory with a central group that understood the full potential of teamwork," he wrote in 2009. "Of course, later, the team unity was reinforced as we stood together on the computer frontier to repel critics.

"There was complete sharing of information. In a bi-weekly report, distributed to everyone, each person reported on progress and difficulties. There was little jockeying for personal advantage.

"Every person has strengths and weaknesses. A team must have a shared vision of the future, a sensitivity to political matters, the capability of developing people, technical competence, the courage to transcend adversity, salesmanship, integrity, the ability to put long-range goals ahead of the short term, and a shared understanding of the individual strengths and weaknesses within the group. We had those characteristics well represented, scattered throughout our group. No person had all these skills. For every person there would be a glaring hole in one or more of those dimensions:

> Yet, it was a group that understood each other well enough to use people in situations where their strengths prevailed and have others compensate for their weaknesses. Out of that came an organization that was able to be much more effective than most of those we see around us in technology and corporations at the present time.

Bob Everett headed the all-important Block Diagrams Group, which would, as he reflected back years later in his "Whirlwind" chapter from *A History of Computing in the Twentieth Century*, "ascertain machine computing techniques, programming techniques, and component designs for accomplishing computing, storing, switching, and programming." Two more bright boys, Stephen Dodd and Pat Youtz, would work part-time with Everett. Other bright boys enlisted to the cause included Harris Fahnestock, in charge of Administration, and David Brown, to lead Electronic Engineering.

Putting the simplest things down on paper helps to visualize them; putting a highly complex, never-before-attempted machine on paper was a must. "We were engaged in an endeavor," said Forrester, "that no one else could claim to understand."

Everett's Block Diagrams Group worked its way through the tricky aspects of knitting all the essential components of Whirlwind together before committing anything to assembly. Some things were simple enough, like figuring out the need for standard electronic and relay racks accommodating removable bases 17 inches by 10 inches. Sylvania was nearby and could easily fill the need.

Other things were more subtle, and the Diagrams Group let out a collective sigh of relief each time a dangerous one was encountered and not yet built. Such was the case when first diagramming Whirlwind as a serial computer (serial computations are done one at a time in succession) as opposed to a parallel computer where more than one computation can take place simultaneously.

"Block diagrams that Everett had developed for serial transmission—inspired originally by the EDVAC machine—had convinced the engineers that, despite relatively simple and easy-to-maintain circuits, such a device

PROJECT
WHIRLWIND
(DEVICE 24-X-3)

REPORT R-127
WHIRLWIND I COMPUTER
BLOCK DIAGRAMS

Submitted to the
SPECIAL DEVICES CENTER
OFFICE OF NAVAL RESEARCH
Under Contract N5ori60

Report by
R. R. Everett & F. E. Swain
SERVOMECHANISMS LABORATORY
MASSACHUSETTS INSTITUTE OF TECHNOLOGY
Cambridge 39, Massachusetts

Project DIC 6345
September 4, 1947

Front cover of Report R-127, 1947.
(Courtesy of MIT Archives.)

would be slow." Too slow for the real-time computer that they intended Whirlwind to be. Much more diagramming and re-diagramming took place, and many more sighs of relief issued up from the bright boys as they plodded along from tube to connector to circuit to voltage.

"By 1947, the logical design of Whirlwind had been completed." On September 4, 1947, the 132-page *Report R-127, Whirlwind I Computer Block Diagrams*, was submitted to the Navy's Special Devices Center.

They were cautious in the design of Whirlwind; it was a first-ever. Forrester and Everett knew how careful Presper Eckert had been with every aspect of ENIAC. Eckert had even tested electrical wire coverings on a cage full of Moore School rats to see which they favored most and least. Little things can mean everything when working on an electronic computer.

If anything goes wrong, even a seemingly small mistake can be disastrous, and may be all but impossible to find and correct. Forrester argued that taking precautions took a "large part of our time and cost." He noted that unlike a radio that "could accommodate some static, a television some flicker, and a teletype some error without destroying a message … a digital computer could tolerate no such error without destroying the message … a single error could produce chaos."

In 1947, while still diagramming their way through *Report R-127,* Forrester and Everett scurried up a bit of their spare time to churn out some extended reading matter for the Navy on the military uses of computers and radar. They wanted to reinforce for the Navy Whirlwind's utility and long-term value, that the Navy's money had been wisely invested in their computer.

Known as *Servomechanisms Laboratory Reports L-1* and *L-2*, which to the Navy must have read like Jules Verne takes on the 6th Fleet, they contemplated how Whirlwind would assist in "a problem involving 10 ships,

5 submarines, interconnecting radar and sonar data, and depth charges in any number up to 20 pre-set units and 20 proximity-fuse units in the water at one time."

Report L-1 concerned itself mainly with Whirlwind helping a destroyer to acquire target data for depth charges; *Report L-2* dealt with combat command and control information and communication among ships of an antisubmarine task group. The reports not only broke new ground for computers in general but also discussed topics totally unheard of anywhere prior to *Reports L-1* and *L-2*.

On paper, the bright boys were visualizing for their ONR benefactors the potential power and versatility of a computing machine that went way beyond that of simply crunching numbers. For Forrester and Everett, "to the best of their knowledge at the time and in after years, [they] knew of no other practical engineering work on how the logic of computers could be applied to interpret radar data." For the military, this view of Whirlwind and computing—there in 1947!—was eye popping.

A Place Called Home: The Barta Building
In 1948 Forester, Everett, their band of bright boys, and 132 pages of Whirlwind's block diagrams moved from Gordon Brown's Servomechanisms Laboratory into the Barta Building at 211 Massachusetts Avenue, a block from the front door of MIT, just around the corner from John Marchetti's Air Force Cambridge Research Laboratory at 224 Albany Street, and a quick jaunt from George Valley's office in the Physics Department.

On the front door a sign now read, "The Digital Computer Laboratory," and Forrester was its director.

Cambridge city records state that E&R Laundry, an industrial cleaner, put the building up. That's probably why it had a 60-foot smokestack out back. Still there but in disuse when the boys arrived, the smokestack's massive base is all that remains today; its ancient iron hatch

Exterior of Barta Building, 211 Massachusetts Avenue, circa 1949. (Courtesy of MITRE Corporation.)

inscribed in Latin boasts that the furnace was made in New York City at the turn of the century.

Rusticated concrete windows and corner trim set off and brightly contrast with the dark exterior of red brick. It is said that the beaux-arts style of the Chicago World's Fair of 1893 inspired the building's design. A grand cartouche etched with the year 1904 is perched majestically above the massive oak door.

Ornate escutcheons, each inscribed with a large floral "B," and separated like belt loops around the girth of the two-story structure, came later with the building's second owner, the printer Barta Press, which sold it to MIT. A bit grimy and worse for the wear after sitting on the sidewalk for 44 years, the building's original sheen had long since tarnished by 1948. But it still had charm. Any bright boy needing to get away from the others for a bit of seclusion could nip off to the charming, rooftop turret with its copper cupola.

Beneath the turret a fancy trim of ornamental molding ringed the second floor; molding interspersed by small gargoyles with faces like theatrical masks...some smiling, some frowning, some angry, some crying, some laughing, and some aghast—every human emotion at the ready—to bear witness to whatever performance the boys had in store.

What a wonderful place for a computer to grow up. And in the spring of 1948, the bright boys began to raise their offspring there.

During the move, they jettisoned their work on the flight analyzer (officially, the Aircraft Stability and Control Analyzer). It was a major part of their contract with the Office of Naval Research, and their decision did not sit well at all with the ONR.

It had not seemed too rash a decision at the time. After all, building an electronic digital computer, as they had witnessed at the Moore School, was no trivial task. Pursuing a single machine, the more important and practical of the two, made real sense.

The problem was that hardly anyone realized how difficult it was to build an electronic digital computer. Forrester and Everett were convinced that simultaneously trying to build both computer and flight analyzer might have jeopardized both.

Redmond and Smith note (*Project Whirlwind: A Case History in Contemporary Technology*) that the Navy was well aware that the flight analyzer might be dropped. In 1946, in an exchange of letters between Forrester and Lt. Commander K. C. Knutson of the ONR's Special Devices

Division that predates the Moore School lectures, Knutson had asked Forrester about "applications of high-speed, electronic computation" applied to naval operations.

Forrester replied with a detailed description of Whirlwind's capabilities serving a "coordinated Combat Information Center … for rocket and guided missile warfare … automatic radar tracking and fire control … aerial and submarine torpedoes … [and the] stability and trajectories of guided missiles." Forrester had already eliminated the flight analyzer from the equation.

"The Navy Special Devices engineers," wrote Redmond and Smith, "enthusiastically realized that they were contemplating a revolutionary device … "

Unfortunately, the Special Devices Division was diminished into the Special Devices Center, and soon after, its control over computers was transferred to the ONR's Mathematics Branch, headed by Mina Rees, since its beginning in 1946. The fate of Forrester's vision for a Combat Information Center disappeared in the shuffle.

By 1947, the decision was full bore ahead with the behemoth computer only, which they reckoned was plenty enough of a challenge. However, eliminating the flight analyzer had its consequences, because it effectively eliminated their ONR mission as well as a Navy contract stipulation about building a computer-controlled flight analyzer.

Without a defined mission, Whirlwind was reclassified by the ONR and placed under supervision of Rees' Mathematics Branch (reconstituted as the Mathematical Sciences Division in 1949), which meant that, like ENIAC, it would fall under the sway of mathematicians, with their emphasis on pure and applied mathematics, statistics, and numerical analysis.

Charles V. L. Smith, former engineer and section head at Raytheon, skippered the computer program. In 1949, with the advent of the Mathematical Sciences Division, Smith became leader of its newly formed Computer Branch. In February of the same year, mathematician Mina Rees was again promoted, this time appointed as director of the new Mathematical Sciences Division.

None of this augured well for the bright boys and their newfangled, electronic, real-time, general-purpose, digital machine. Whirlwind became just one of the dozen or so other computers under the control of the Mathematics Branch and then the Mathematical Sciences Division. Rees remarked, sight unseen about Whirlwind: "too much talk and not enough machine."

From February 1948 onward, time schedules were drawn up and the layout of the computer began. The bright boys kept monthly Summary Reports that kept the ONR posted on every aspect of their progress. Forrester intended that the reports be so thorough and exacting that they would make a "virtue of their frankness" such that the ONR would never call into question Whirlwind's forward progress. They had the opposite effect. The reports showed plainly how the bright boys went about things: too thorough, too painstaking, and too expensive in the ONR's estimation.

Things backfired even worse when the now overly curious ONR sent agents to check on the activities in the Barta Building.

Hardly a month went by when someone or other from the Office of Naval Research in Washington, the ONR's local Boston office, or someone else at the request of the ONR's Boston or Washington offices, or from MIT itself, did not sweep into the old laundry building to call into question the budget, the plans, the personnel, the materials used, the project's mission, its ultimate utility, or its completion date.

Dropping the flight analyzer had produced immediate, unforeseen, and very uncomfortable consequences.

Analog MIT

The bright boys had no competition anywhere at MIT, or uptown at Harvard or anywhere else in Cambridge, or Boston, for that matter, or anywhere south of the Charles River right to the doors of the Whistle Factory.

Although they had yet to build out even the barest of structure for their electronic digital computer, they were far ahead of the outdated analog machines around them.

The Bush Differential Analyzers from the 1930s were still at MIT, as was the Rockefeller Differential Analyzer (an attempt to make an electronic version of a Bush Analyzer), plus IBM 604 and 602 calculators, punched-card machinery, and some engineering plans from Wilcox Overbeck for another analog device to be called the Rapid Arithmetical Machine, which never got built. Work on the Rapid Arithmetical Machine was halted in 1942. With their return from the Moore School, Forrester and Everett were convinced more than ever that a remarkable opportunity had been set before them at MIT to build a very extraordinary type of computing machine.

The big difference in their favor was that they intended to build a control computer—a machine that would automatically control the actions of yet another machine. Early on, that other machine was a flight control analyzer. Seemingly everyone else building computers had a fixation on constructing single-purpose, scientific computers used solely to hurry into being answers to previously intractable mathematics problems.

However, replacing Whirlwind's flight analyzer/simulator with a different machine, like one to operate a skyscraper's lighting system or another to open the spillway at Hoover Dam, and the computer alters their reality: lights pop on and off in the skyscraper and millions of tons of water cascade out of a mammoth dam. Place Whirlwind into the practical hands of engineers surrounded by countless other machines, and the possibilities of linkage begin to blossom. Although when starting out, Forrester and Everett did not have an inclination toward controlling anything other than a flight simulator/analyzer, the skids were already greased and waiting.

MIT's kingdom of gears was called the Center for Analysis and was presided over by Samuel Caldwell, colleague and partner with Vannevar Bush in the making of the Differential Analyzer. Although in the summer of 1945 the Navy offered to finance the construction of an electronic digital machine if the center would submit a plan for one, Caldwell declined. It was a good decision because neither MIT nor Caldwell had the capacity to pull off such a request.

Although MIT vice president James Killian reported to the school's president, Karl Compton, that the center was still a going concern and had "contacts, reserve funds, prestige, and staff" to be a key player, it really did not have the foggiest idea of what was afoot in electronic digital computing. The only true shot at MIT producing such a machine was with the bright boys. As Forrester repeated many times: "We were engaged in an endeavor that no one else could claim to understand."

While Sam Caldwell listened to gears grinding to a halt around him, postwar aftershocks were rumbling beneath the venerable foundation of MIT's School of Engineering. The wartime work of the Rad Lab, dominated by physicists, had exposed the electrical engineering curriculum as being as backward as Caldwell's gear factory. Stuck in the mold of the classic electrical laboratory with a mindset, expertise, and tools for large-scale, electrical power generation, MIT engineering was in dire need of serious revamping toward electronics, electromagnetic theory, communications, and circuit theory.

Harold Hazen, dean of the department, said rather bitterly—while at the same time liberally spreading the blame around—that "[e]ngineering over the country as a whole did not have the intellectual standing that science had." Of course, intellectual standing had little to do with the situation. A dated curriculum has no intellect; it is just out of touch with reality.

Then too, this was also a case of the boys having at each other again. It was the same mewling dogfight between engineers and physicists that went on during World War I when Thomas Edison refused the American Physical Society (physicists) and the National Academy of Science from participating in war research through his Naval Consulting Board.

The reason given by Edison's chief engineer, M. R. Hutchinson, was that the Naval Consulting Board should have a membership "of practical men who are accustomed to doing things, and not talking about it."

The spat ended when George Ellery Hale, director of the Mount Wilson Observatory, eventually persuaded President Woodrow Wilson to establish the separate but equal National Research Council for scientific contributions to the war effort.

Hazen had good reason to be so defensive; the Rad Lab's physicists had blown the doors off most everything with their pulse circuits, feedback, and microwaves. Their astounding displays made glaring the inadequacies of engineering.

Karl Compton, as a physicist and MIT president, well understood such situations. He had experienced two of them—once from his days as a Princeton physicist while reconstructing its Physics Department, and then again at MIT, when he hired John Slater, together with Philip Morse, Wayne Nottingham, and Robert van de Graaff, to elevate MIT physics.

During the Slater-led transformation, Compton had eased out the old physics chairman and five senior faculty members to help the process of change run more smoothly.

He was again faced with a similar challenge, this time in the field of electrical engineering. Engineers like Forrester, Everett, Brown, and the entire Servomechanisms Lab were already well indoctrinated into the wonders of electronics from their stints working Rad Lab projects. J. Presper Eckert had likewise, learning about the new arts in roughly the same manner, having done wartime contract work for the Rad Lab at the Moore School; so too had William Shockley (co-inventor of the transistor in 1948) gotten similar exposure through his invention of the mercury delay line for Rad Lab radar.

It was getting quite evident that electrical engineers without such exposure were way behind the eight ball. Sam Caldwell was one of them, and all he ever needed to get caught up was to catch the ear of a very able *digeratus* in his own Center for Analysis, one Perry Crawford, who had been the first to steer Forrester toward electronics. Caldwell's signature even adorns Perry Crawford's master's thesis.

Historian Larry Owens points out from correspondence between MIT's provost Julius Stratton and then-president James Killian that it was common knowledge at MIT that the postwar Center for Analysis had lost the confidence of other departments and had begun visibly to crumble. As John Maynard Keynes observed wryly, "The difficulty lies not in the new ideas, but in escaping the old ones."

If the war had speedily transformed cloth-covered biplanes into metal monoplanes then to swept-back wings and jet engines, then what the heck had Sam Caldwell been doing all the while? Was he under a rock or at the controls of a bygone era flying in circles?

Caldwell flirted with success in 1946 when MIT applied for and received a $100,000 grant from the Rockefeller Foundation, not to build a computer "but rather to study how to bring digital and electronic technologies into MIT's computing program." In terms of 1946 dollars that was a helluva rich offer for a technology that the bright boys were already bringing to the university. But, who's to argue with a tidy and very timely grant for the Center for Analysis?

The center's lifeblood seemed to be grants; it was established with a grant from the Carnegie Corporation to Harold Hazen's Electrical Engineering Department, then sustained by a Rockefeller grant to build an electronic version of the Bush Differential Analyzer. MIT returned part of the grant when the conversion faltered and was stopped.

But in 1946, Caldwell and MIT were back at the Rockefeller money trough for more, and got more. The carrot in the deal for the $100,000 was the intimation that electronic computation at the Center for Analysis might lure MIT's resident math genius, Norbert Wiener, to set up shop there. It didn't happen; Wiener showed no interest and went off to Mexico to pursue the intellectual trajectory of his cybernetics research. Thereafter, Caldwell's time and that of the Center for Analysis grew short. MIT chose not to put any money or effort into sustaining the moribund facility.

The center and Caldwell sank quickly and disappeared from sight. It took a few years for MIT to swallow such a defeat, but in "1950 MIT

Provost Julius Stratton confessed to president Killian that MIT had muffed its chances to establish leadership in computing." Phoenix-like, it arose again nearly 10 years later, in 1957, as the Center for Machine Computation with Phil Morse, a physicist, at its helm.

Postwar aftershocks also unsettled the way MIT financed its basic research operations. Going forward, Whirlwind was a good future look at the coming research model: military projects done with military money for military technology that may eventually in some manner, shape, or form go commercial.

The prewar days of philanthropic gifts, grants, and industrial funding alone pulling the research wagon along were gone; any thought of returning to them was an idealistic vision that would "turn out to be an endlessly receding mirage." The world had big problems that required big science to fix, which needed real big money that only the government's wallet could cover.

Whirlwind's benefactor, the Office of Naval Research, took the lead in funding university science and electronics. "Bolstered by millions of dollars from recently cancelled procurement contracts, the ONR became the predominant patron of academic research, and set the precedent for postwar relationships between the military and the university."

The Army was not far behind. A memo—more like a veiled command—from General Eisenhower, dated April 30, 1946, was given to Compton and Stratton:

> The lessons of the last war are clear. The armed forces could not have won the war alone. Scientists and businessmen contributed techniques and weapons which enabled us to outwit and overwhelm the enemy.
>
> Their understanding of the Army's needs made possible the highest degree of cooperation. This pattern of integration must be translated into a peacetime counterpart which will not merely familiarize the Army with the progress made in science and industry, but draw into our planning for national security all the civilian resources which can contribute to the defense of the country.

The Buildout

Forrester and Everett insisted that all parts, assemblies, and subassemblies be tested individually and then together, and that designers should take the time to build whatever test equipment was necessary to locate deficiencies and remedy them before building them into Whirlwind.

And build nearly everything they must; they couldn't nip out to the nearest electronic supply house for their computer gear—there weren't any such supply shops or any such gear. They had to rummage for parts or have them specially designed and built.

Everything about Whirlwind was new, untried, and difficult. And very quickly the bright boys found themselves dealing with a previously unheard-of level of complexity—pioneering it. It was like a newfound land to colonize and civilize.

One false step and they would be lost. Only their brethren from the Moore School lectures, off on their own somewhere hacking through similar dense stands of the unknown, could appreciate their plight.

Gordon Brown, their boss in

Interior of Barta Building and Whirlwind memory unit: Charles Corderman (on ladder); Gus O'Brien (below left); and Norm Daggett (below right). (Courtesy of MITRE Corporation.)

the Servomechanisms Laboratory, realized it, and his benign neglect gave his bright boys time and space for their thoughts to rattle about and gain enough composure to come through the experience. "We saw Gordon infrequently, maybe once a year," Everett later recalled, warmly thankful for Brown's hands-off approach. "Otherwise he left us to our own devices."

Complexity not only engaged them with their machine and its intricate relationship of parts and functions, but also dwelled within each of them.

Immersed in complexity as they were, they began to think in complex patterns and share complex notions.

They peppered their engineering journals with the process: self-doubt, false starts, and then hurrahs of delight at every success, even the smallest, before finally the new knowledge settled in for good and became a part of them. "There are vast realms of consciousness still undreamed of," said D. H. Lawrence in "Terra Incognita," his homage to the unknown, "vast ranges of experience, like the humming of unseen harps, we know nothing of, within us."

Constantly thinking about complexity, the myriad parts of their machine, its systems and subsystems, the changing relationships among the parts as an electrical force was applied to them, and the resulting actions and reactions taking place within and among the system because of that force, opened doors of insight and perception that were as new and powerful to them as their new machine.

And as Lawrence concludes, like the "slow mounting marvel of a little tree," complexity began to reveal itself, bit by bit taking up residence in the intellect with its newfound enlightenment.

The bright boys might not have had standing with senior academicians or have authored scores of papers and textbooks, but they now possessed a deep knowledge and skill the likes of which no one could challenge.

Better yet, they knew how to put that knowledge to work. They had a tiger by the tail and were riding it for all it was worth. "We were cocky," reminisced a bright boy years later. "Oh, we were cocky! We were going to show everybody. And we did."

And only the best would do for them as they transformed their patch of complexity from *incognita* to *cognita*. They felt entitled to their seeming budgetary extravagance, saying that it "was a deliberate policy of saving development time and money in the long run by insisting on going first class." Many of the bright boys had witnessed the very same extravagance working its magic at the Rad Lab during the war. The Rad Lab environment was an incubator of wonderful discovery … but for a price.

By September 1948, even MIT began to take notice. The bright boys were converting mathematical, logical, and abstract concepts into actual working machinery. Trucks were arriving from Sylvania with the racks that formed the skeletal outline of the computer's superstructure.

Clusters of assemblies and subassemblies of components and subcomponents began to flesh out Whirlwind's bones. MIT leadership "became aware that they had a partly finished, well-begun computer of unique design on their hands." The behemoth was rising.

The bright boys would design, build, test, integrate, redesign, rebuild, retest, and reintegrate. They felt unbound from any set of engineering

requirements, even
their own. If along
the way they hit
upon a better pro-
cess or component,
they jumped at it,
tested it, and then,
if it improved their
design, they inte-
grated it into their
machine.

Schematic drawing of Whirlwind computer layout. (Courtesy of MIT Archives.)

They were young, brilliant, confident, and cocky, which for many hov-
ering nearby waiting impatiently for Whirlwind to go bust, was grandly
irksome. There were remarks that the bright boys were "arrogantly high-
hat and snobbish ... that they were as unrealistic about what they were
doing as they were young and immature, and theirs was a gold-plated
boondoggle, extravagant in its demands, in its rewards, and in its raids
upon the taxpayers' purse."

Such an unexpected discovery came about when dealing with the pesky,
ever-unreliable electron (vacuum) tubes of the day. They happened upon
the problem that caused the tube's short lifespan. The source was the fabrication of the nickel cathode within the tube itself: it was the tube's own
innards that doomed it.

The manufacturer, in order to make the cathodes easier to shape,
included a small amount of silicon in the nickel from which the cathode
was made. After some 500 hours of operation, a high-resistance, monomo-
lecular layer of silicon would migrate to the surface of the cathode and bias
off the flow of electrons.

The problem had not been, as assumed, a failure of the cathode coating
to emit electrons, but instead was internal resistance to the flow of current. The bright boys had tubes built to their own specifications of high-
purity nickel cathodes even though they were somewhat harder to shape.
The result increased the life of vacuum tubes a thousand-fold. Because of
their efforts, Sylvania produced the 7AK7 tube in 1948, which into the late
1950s was the standard tube for all computers.

Such an eye-opening lesson in the bright boys' brand of quality control had a profound reaction at Sylvania. "The Sylvania engineer who
was in charge of making our tubes," recalled Forrester, "realized that

production should be insulated from the past practices of making cheap tubes. He insisted that the tubes be made in a city where tubes had never been made before so that bad practices did not have to be unlearned." As Forrester concluded, "Very insightful."

Such improvements escaped the ONR's purview and went down simply as line-item expenses—irritatingly intolerable expenses. With the 7AK7 undoubtedly an integral and expensive part of many, maybe all, of the other ONR-sponsored computing projects, a little thanks might have been more appropriate.

The new Chief of Naval Research, Admiral T. A. Solberg approached MIT's president Karl Compton on September 2, 1948, with the news that Project Whirlwind's funds "future commitments and rate of expenditure be scaled down" until such time as both the technical and financial requirements of the project could be reevaluated.

It came at a bad time for Compton. He was getting ready to resign his MIT presidency and go to Washington to replace Vannevar Bush as head of the Research and Development Board (RDB), which had temporarily replaced the Office of Scientific Research and Development (OSRD) until such time as the National Research Foundation would come into being (initially vetoed by Truman in 1947; later signed into law by Truman as the National Science Foundation in 1950).

On September 8, Compton asked Nat Sage to get him up to speed on Whirlwind, and on electronic digital computers generally. He needed the information for two reasons. First, he knew little about the new electronic machines, and second, he would be in a position in Washington to do a little PR for Whirlwind if he had a good document to circulate around.

Nine days later on September 17, Compton had his report. Forrester and Everett, together with Hugh Boyd, Harris Fahnestock, and Robert Nelson, put out *Servomechanisms Memorandum L-3*, titled "Forecast for Military Systems Using Electronic Digital Computers." *L-3* had two parts: a budget proposal covering fiscal years 1949 to 1953 and a 2-by-3-foot foldout page on which there was a grid with a 15-year forecast (1948–1963) of computers in the military, covering guided missile data reduction, cryptography, high-speed computer networks, interception networks, air traffic control, industrial process control, simulation and training, and logistics. Each column in the grid represented a year and each row was dedicated to a future military capability that computers would serve.

L-3 looked like a graphical companion to *L-1* and *L-2*. The practical engineering side of the bright boys could not resist making abundantly clear the potential applications for their new electronic digital wonder machine—some 15 years into the future!

On a broader level the report represented a first-ever call to action for the country to begin a computer development program on a national scale; a claim that it would be a far better alternative than the scattered investments in one-off machines as was the case up to 1948.

By 1948, time and technology had also rendered moot the need for the 1944-era flight control analyzer. Three years on from the original request, the de Florez device was already seriously dated.

Pre-war subsonic aircraft analysis was now supplanted by the transonic and supersonic, with engineering requirements that went far beyond the bright boys' capabilities. The need to display and solve some 90 equations in real time, like aerodynamic coefficients, instrument equations, equations for motion, and earth axes would require a special-purpose computer, and the necessary technology just wasn't anywhere and wouldn't be for some time. The dramatic shift away from the subsonic was quite evident just across campus where MIT's Department of Aeronautics was fast about building competency in transonic and supersonic flight, including, with more Navy money, a new wind tunnel for testing aircraft in the Mach 2 and Mach 4 ranges.

Whatever the bright boys had for a flight analyzer was by 1947 best used for something else; any future for any flight control analyzer would first need serious attention from specialists like aeronautical physicists and flight dynamics engineers. Across campus, MIT's Robert Seamans, much later an executive director at NASA and even later Secretary of the Air Force, directed 120 staff on the Department of Aeronautics' flight dynamics project, which was all about the future of jet aircraft and missiles.

Transonic and supersonic flight analysis demanded high-end accuracy, and Whirlwind could not meet the arithmetic functions needed for high performance analysis of modern airframes.

If Forrester and Everett had stuck with both the digital computer and the flight analyzer, they would have most likely overspent and underproduced for a very long period of time, and in the end might have very probably been overtaken by the flight analyzer competition, which included the all-time heavy hitter, Edwin Link, the father of flight training and analysis.

The Navy would eventually get its special-purpose trainer/analyzer, research for which began in 1950 at the University of Pennsylvania. Ten years later in 1960, the fruit of that research, called the Universal Digital Operational Flight Trainer or UDOFT, rolled out of a Sylvania factory. Soon after the UDOFT appeared, Edwin Link came out with his famous Link Mark 1.

Dropping the flight analyzer was a gamble that paid off. However, it took a lot of forbearance on the part of the Navy to accept major changes to a product on which they expected Forrester and Everett to make good. After all, the Navy had already fronted the bright boys nearly $1 million ($75,000 in December 1944 and then $875,000 more in June 1945) with no flight analyzer to show for the investment. Forrester then sold his Navy patron on going digital in 1946, which got him another fat contract and more time. No wonder that the Navy saw Whirlwind as an embarrassing money pit.

The bright boys did, however, keep the Navy informed at every stage of their research. At any time throughout Whirlwind's birthing, the Navy had complete access to every step of the process. The bright boys risked the Navy's money, everyone's time, and their shot at big-time success because they strove to build reliability into Whirlwind, which was the big knock against all of the early electronic digital computers. Reliability is what chewed up the Navy's investment the most.

The few big machines that were in existence in 1947 were wonderful, but most failed after short runs. Many times, they ground to a halt owing to rather minor malfunctions, which is why the bright boys aimed for Whirlwind to be a near flawless, highly reliable machine.

And in 1947 their quest began to bear fruit. First out of the box was bright boy Norman Taylor who delivered on real-time computation with a small (6 feet high by 15 feet long) testing machine tabbed the 5-digit multiplier. Taylor got the machine to multiply two 5-digit binary numbers in 5 microseconds with high reliability, and soon after got the machine to multiply for 45 days without an error. That many errorless days of reliable operation was simply unheard of from any machine anywhere.

When John von Neumann first visited the bright boys' 5-digit multiplier, it had just completed 5 billion multiplications without an error. He got so excited that he kissed Norman Taylor on both cheeks.

Von Neumann's kiss was less in congratulations for the machine's ability to run without error and more a mathematician's exuberance at the prospect of freeing calculation from the drudgery of doing it by hand or resorting to Howard Aiken's Mark I electromechanical machine.

Norman Taylor at the 5-digit multiplier. (Courtesy of MITRE Corporation.)

Von Neumann recognized in a heartbeat that his calculations should and would only be run on an electronic digital computer. The reason was simple: they were the only machines that could do the job that he needed to have done.

The introduction to electronic computing from ENIAC and now the five-digit multiplier provided von Neumann with the only way to find answers to previously intractable problems. He was more than grateful for the opportunity and hell-bent on electronic computation assisting his work.

"The phenomenon of turbulence, the dynamics of explosions and implosions, and the modeling of complex weather patterns," wrote Howard Rheingold in *Tools for Thought*, "each required such enormous numbers of calculations the future progress in the field was severely limited by the human inability to calculate the results of the most interesting equations in a reasonable length of time:

> With an electronic computer plowing through equations for him, von Neumann could now arrive at a clearer understanding of the forces at work in these phenomena.

He resolved to build his own at Princeton's Institute for Advanced Study (IAS). With funding from the U.S. Army and Naval Ordnance, and help from ENIAC veterans Arthur Burks and Herman Goldstine, as well as cybernetics pioneer Julian Bigelow, he built an electronic, real-time computer known simply as the IAS computer.

Sooner than later, the ONR's Computer Branch started looking at Whirlwind and the IAS computer as equals—except for the steep price tag that came with Whirlwind.

Fragile Memory

In addition to electron tubes as on/off switches for logic circuits, there were also giant cousins, called electrostatic storage tubes, which were used for storing binary data, what we now refer to as RAM or random access memory. In 1947, these tubes were considered state of the art, predating the transistor, which was invented in 1948, but barely employed in computers until 1953.

These electron memory tubes replaced the older and much slower technology of the Mercury Delay Line. A delay line was a 5-foot tube of mercury (usually used in sets of tubes and referred to as lines) in which an electric pulse, representing a binary bit, was converted into a sound pulse that then traveled as a sound wave through the tube of mercury, emerging at the other end where it was reconverted into an electric pulse again.

About a thousand binary bits could be stored in each mercury tube, but accessing that memory was slow—hence, the choice of an electron tube which was much faster. Today, of course, both electron tubes and mercury lines are long gone, replaced at first by integrated circuits or ICs populated by teeny transistors that acted as the on/off switches and RAM, and later still, the ICs were supplanted by microprocessors

Whirlwind's on-site tube manufacturing department: (left to right)—Pat Youtz, Stephen Dodd, and Jay Forrester examine the finished electrostatic storage tube, 1951. (Courtesy of MITRE Corporation.)

inhabited by millions of transistors, whose transistor populations seem to grow by the millions every year.

The problems with the electron tube switch as well as electrostatic tube memory were that they were expensive—especially electrostatic memory tubes—short lived, and could blow out unexpectedly.

Maintaining the reliability of electron tube switches, numbering in the thousands in a computer, was a frustratingly nasty affair. They aged quickly and badly; they leaked electrons, lost efficiency and then burned out. Burnouts caused computer errors that would shut a machine down until the tubes were replaced.

"Radio engineers were not concerned that the life of a vacuum [electron] tube was about 500 hours," noted Forrester. "But computer engineers, considering the use of many thousands of vacuum tubes [Whirlwind had 5,000 electron tube on/off switches], easily estimated that with such a short life, the machine would run no more than a few minutes between failures."

The much larger, electrostatic storage tubes, either commercially available or those custom built by bright boys Stephen Dodd and Pat Youtz, had extraordinarily brief life spans and "were costing us about a thousand dollars [each] to make," recalled Forrester. "They would store about a thousand binary digits and last about a month."

Resolving tube problems while fending off pressure from their frequent interrogators took considerable tact and calm. But they came through it. To counter imminent tube failure that would have rendered a computation utterly useless took a Forrester brainstorm to solve. Using Taylor's machine, Forrester invented marginal checking, which automatically spotted the tubes that were soon to fail. Replacing those tubes early on gave them the fast switching and dependable memory that they sought for Whirlwind.

Forrester realized, however, that marginal checking of electrostatic memory was merely a stopgap, that something more permanent would soon be necessary. He began a search for such a memory capability, and that quest would eventually lead him to one of the most important discoveries in computer history: magnetic-core memory.

For bright boy Dudley Buck, Whirlwind's gas-filled, glass tubes and glowing filaments would launch the 23-year-old research assistant into the chilly world of superconductivity, which is the ability of some metals to conduct electric current with no resistance at extremely low temperatures, like those below minus 420°F. Buck's eventual discovery of the cryotron, a superconductive computer switch, based on the effects of magnetic fields at liquid helium temperatures, represented the first practical use of superconductivity.

A hundred cryotron switches could fit inside a thimble, a far cry from the space needed by, say, a hundred electron tubes. Although Whirlwind never wore any of Buck's cryotrons, just the machine's presence on campus had an alluring power to grab a mind like Buck's and to switch it on.

In the years to come, Whirlwind would switch on many other minds. Robert Noyce for one, who in 1947 was still an undergraduate at Grinnell College in Iowa. He arrived at MIT in 1949 as a graduate student hell-bent with taking on electronic digital computers, and he would soon after co-invent the integrated circuit and then go on to co-found the Intel Corporation.

A youthful Ken Olsen prowled the monstrous racks of Whirlwind and toted away the formula for a modular computer that became the hallmark of his equally monstrous computer company of the 1980s, Digital Equipment Corporation, or DEC.

There were indeed other worlds within the machine, *terrae incognitae*, "vast ranges of experience" yet to be discovered that were as important as the machine itself. And Forrester and Everett were its gatekeepers. Technology and fortunes beyond all imagining shimmered between the covers of *Report R-127*.

In 1947, the long-range hope for Whirlwind as a real-time, general-purpose computer rested with *L-1* and *L-2*, and the hope that those reports would generate some attention and recognition in the right places. The bright boys' direction for Whirlwind was concerned with early warning, detection, interception, and command and control above and beneath the water, with sonar and radar data and communications all tied together. Such potential would have made for joyful pandemonium in the halls of the Pentagon, if the news ever reached the right ears.

Whirlwind was being crafted into the brain of an information system capable of keeping a close eye on the movements of, say, Stalin's hordes, or maybe of providing a bit of strategic forewarning about an impending attack. *L-1* and *L-2* might also divert for Whirlwind some of that rich flow of research money that seemed to be endlessly cascading into MIT from the military.

That early promise for *L-1* and *L-2* withered in 1947. The ONR missed a huge chance at shoring up a good chunk of national security. And it was exactly the kind of national protection that Congress had in mind when it created the Office of Naval Research in 1946: "to plan, foster, and encourage scientific research in recognition of its paramount importance... for the preservation of national security." And when over time the Special Devices Center was absorbed by the newly formed Computer Branch (1949) of the Mathematical Sciences Division, and soon thereafter, as mathematicians from the Computer Branch took up scrutiny of Project Whirlwind, the cocky, mission-jeopardizing bright boys were in the hot seat for getting rid of the flight analyzer.

The Budget

Twenty days after the previous ONR budget meeting, MIT met again with the ONR on September 22, 1948. It was a serious meeting that detailed the impending financial crunch for Whirlwind for fiscal year 1949, which would end on June 30, 1949. Forrester's request for $150,000 per month

($1,800,000 for the year) was met by the ONR's firm counter of $900,000 (or $75,000 per month). A draconian cutback for Whirlwind.

The machine's monthly glass bill alone would eat up a huge chunk of that. Whirlwind's 32 electrostatic memory tubes were costing $32,000 a month. Whirlwind was big and expensive, and none of it was news to anyone, especially the ONR. The computer was building out to cover an incredible 2,500 square feet of floor space—two and a half times larger than ENIAC. The ONR offer was not killing the beast outright; it was squeezing it into extinction by the month.

Nat Sage offered a way out, saying that $1,200,000 would do the trick for fiscal year 1949, and that a carryover from fiscal year 1948 would up the ante to about $132,000 monthly in working capital for the upcoming year. Relenting a bit during the meeting, the ONR conceded that more money would be forthcoming if Whirlwind seemed to be making and keeping a good faith effort toward the $900,000.

Flamm suggests in *Creating the Computer: Government, Industry, and High Technology* that there were other motives afoot at the meeting, unstated ones like the ONR's sensitivity about the "large fraction of public research funds for computers going to MIT."

Then too, was the ONR ever going to see MIT kick in a few bucks toward Whirlwind's monthly meal ticket? Whirlwind stopped having a Navy mission when the flight analyzer was dropped, yet the ONR was still paying top dollar as though the mission was still ongoing.

Whirlwind was the only electronic digital computer that MIT had on campus and would undoubtedly go to enriching the educations of thousands of MIT students. Did any of that digital advantage count toward any financial consideration on MIT's part? Maybe like offering to split the difference in the contested $75,000 per month?

Nothing of the sort was forthcoming from MIT. With some 200 universities participating in the ONR's work programs, no university or college would be silly enough to suggest sharing expenses. That could well turn out to be a dangerous precedent. Besides, the ONR put itself out as the Daddy Warbucks of research, funding 40 percent of all basic research in the United States and supporting nearly half of all doctoral students in the physical sciences.

In support of computing, the ONR's share was almost total, and would continue to be right to the late 1950s. However, if the ONR cut anyone off from funds or squeezed a project, there was no recourse. You took the hit and moved on to something else.

Building a computer under the ONR's auspices was precarious. The bright boys got their $1,200,000 for fiscal year 1949, but for fiscal year 1950, their budget was slashed to $750,000. That gave them nearly a year, until June 30, 1949, to come up with a plan for their deliverance.

Chapter Five
1949

The Math Problem

In early January 1949, Jay Forrester rendezvoused with Perry Crawford in Washington, DC. Escaping Cambridge, getting away from assaults on Whirlwind's budget and technology to meet with a like-minded digital pioneer and good friend, was a bit of a vacation... no matter how brief.

Forrester and Bob Everett had spent the previous October dashing out another defense of Whirlwind to counter the ONR's claim that Whirlwind and von Neumann's IAS computer were mostly identical machines. The implication: why fund both? An ONR claim that even von Neumann refuted as being without merit.

Known simply as the IAS machine, von Neumann's computer was not real time, because it didn't need to be; and it wasn't general purpose, because it needed to be much more precise.

He was a mathematician probing complex phenomena such as turbulence or the dynamics of explosions and implosions, which demanded a specialized, purpose-built, very precise computer capable of producing results in a reasonable amount of time. Accuracy and precision were absolute, real time need not apply.

The special-purpose IAS machine was not the proper tool for, say, air traffic control, but Whirlwind was. Von Neumann's machine would not be operational until 1952, three years after Whirlwind cranked into action.

Nevertheless, a "Comparison between the Computer Problems at the Institute for Advanced Study and the MIT

Jay Forrester in the Whirlwind control room for a TV interview (c. 1951). (Courtesy of MITRE Corporation.)

Servomechanisms Laboratory" was produced and shipped off to the ONR for review and consideration.

"Jay, my good friend," smiled Perry Crawford good-humoredly, as he listened to Forrester tick off his latest travails with the ONR, "I think you have a math problem."

"Indeed, ongoing for more than a year now," replied Forrester. "Almost monthly we receive a visit from another ONR inquisitor of some sort, who invariably returns to the ONR with an inaccurate assessment of Whirlwind that's cheered as fact, and then reacted to as another instance of our, I guess, incompetence."

"Is it beside the point that Whirlwind works?" scoffed Crawford. "That it's a first-ever, a one of a kind?"

"*Beside the point* is exactly how it has been taken by many," answered Forrester. "As you are well aware, Perry, we've had no Navy mission ever since we jettisoned the analyzer and cockpit. We're now just one of a dozen or so machines financed by the Navy. How unique a machine we've built and how good was our decision to go digital is irrelevant."

"And I feel responsible. I helped steer you into going digital."

"Glad you did," Forrester said with a nod of relief. "We lost an analog mission that was growing more impossible to solve on a daily basis, but replaced it with a calling. No, Perry, you saved the project by giving it new life."

Perry Crawford, 32 years old in 1949, graduated from MIT in 1939, the same year that Jay Forrester first arrived in Cambridge from Nebraska. While a grad student in Sam Caldwell's Center for Analysis, Crawford had written his 1942 master's thesis on using a digital device as a control system.

It was an eye-opening thesis to none other than J. Presper Eckert, co-builder of ENIAC. Eckert, years later in an interview with Christopher Evans, said he had read the young grad student's paper and spotted Crawford's memory device: "I had gotten the idea of using disks for memory, digital memory, from a master's thesis written by Perry Crawford at MIT. He had not built any such disks; it was just speculation."

Eckert never built the device, but its talented author soon thereafter would meet up with Forrester, someone who would very much need a helping hand with computer memory.

By October 1945, a month following Forrester's meeting with Gordon Brown and his selection of the ASCA project, Crawford began working for

Luis de Florez at the Navy's Special Devices Division. As luck would have it, one of Crawford's new duties was as the Navy evaluator for what would become Project Whirlwind.

When apprised of the difficulties that Forrester and Everett were encountering trying to build ASCA, it was Crawford who suggested going digital. That same October, Forrester and Crawford together attended the *Conference on Advanced Computation Techniques* at MIT where they first learned about ENIAC. Later, at the Moore School lectures in 1946, Crawford would deliver one of the presentations: applications of digital computation.

Crawford and Forrester seemed to be perfect digital mates; Crawford even assisted Forrester in convincing the Special Devices Division to agree to separate the analyzer from the project, and then to recast the computer as a digital machine. In his foreword to *Bright Boys*, Forrester remembers Crawford as "a person with continually unfolding visions of futures that others had not yet glimpsed. He was always looking, listening, and projecting new ideas into the future."

If the early days of digital computing needed a guardian angel, Perry Crawford it seems might well have been an ideal candidate.

"We've also been handed a new budget that's less than favorable for the continuation of Project Whirlwind," winced Forrester, uneasily recalling the September budget cuts. "The new fiscal year on June 30, we'll be cut back by 60 percent."

"You've endured a season of grim news," replied Crawford, shocked by Forrester's tale and concerned for his friend.

"There's enough bad news to go around, that's for sure," agreed Forrester. "But we do have a glimmer of hope. We have an air traffic control proposal in with I. J. Gabelman, program manager at the Air Force's Watson Labs. His decision is due soon. It's for two years at $9,000 per month. Doesn't even cover our glass bill for memory tubes, but it's a perfect fit for what we do best. And, it's good to win one every once in a while."

"A timely godsend for Whirlwind," smiled Crawford.

"For all of us, *if* we get the call."

"And your Whirlwind team is up to the job?" asked Crawford.

"Most definitely," Forrester replied quickly. "The *L-1* and *L-2* reports that Bob Everett and I put together back in 1947 speak volumes to our ability at controlling aircraft."

"I've seen *L-3*," added Crawford, smiling into Forrester's look of surprise.

"We did that one for Compton," said Forrester.

"I know," replied Crawford. "Truman appointed him last fall to succeed Vannevar Bush to head the Research and Development Board. Which, by the way, is where I work these days. I'm on the RDB's Ad Hoc Committee on Air Defense."

"Air defense connects with *L-3* very well," smiled Forrester, very pleased with the news of his position on the Committee on Air Defense.

"Perfect title," added Crawford: "Forecast for Military Systems using Electronic Digital Computers." And you're right, it very much connects well with air defense. More intriguing is that it goes out 15 years, 1949–1963. That's quite a reach into the future. Astounding, actually."

"It was meant to be ... by design. Did it have the right effect on the RDB?"

"Direct hit," replied Crawford. "The Air Force has a half dozen or so different problem areas all relating to electronics, potential solutions for which, I think, are right in your wheelhouse."

"Perry, once again it seems that you just might be in the right place at the right time for Whirlwind. It's a bit uncanny, don't you think?" continued Forrester. "You've been an integral part of this project since when, 1945?"

"We're like-minded; we're both digital advocates," answered Crawford. "The theories from my master's thesis you put into practice. I'm grateful for your belief in me. Besides, you're the only one at MIT willing to take the chance."

"We were at an impasse, and you gave us a way forward," returned Forrester. "And the proof is, it's working out for the better."

"Except for a budget bind that's giving you fits."

"Building reliability into something that's never been done before is what's expensive; not dropping ASCA. Digital electronics is a harsh teacher that won't tolerate any missteps."

"Well, the Air Force has come to realize," explained Crawford, "that human identification and plotting of jet aircraft are too slow. The Air Force needs what the generals are vaguely calling an 'automatic interceptor Director System,' which to them is some sort of an electronic system that could somehow automatically vector jet fighters to their targets."

"*L-3*," shot Forrester, as if making a punctuation point to Crawford's description.

"Yes, *L-3* is quite prescient. Going forward, I bet we'll see a marked increase in the need for air intercept and air traffic control."

Forrester returned to Cambridge a bit buoyed from his meet-up with Perry Crawford. As a small victory in the face of a season of bad news,

Watson Labs awarded the contract to Project Whirlwind, besting the likes of Raytheon, the Franklin Institute, and the Teleregister Company (Western Union) in the process.

The contract was assigned to the Applications Study Group of Gordon Welchman (supervisor) together with four other bright boys (Harris Fahnestock, C. Robert Wieser, William Linvill, and Theodore Hildebrandt).

Crawford's RDB Committee on Air Defense would hold a meeting in March, and he secured an invitation—a foot in the door—for the bright boys to send a representative.

Auguries of impending doom mingling with the tantalizing potential for success flitted back and forth during the whole of 1949. Not knowing which would strike first, they forged on, building out their machine, stretching it out on either side of a long corridor, eight racks deep and 4,000 electron tubes removed from its central control room.

Interior of the Barta Building with the Whirlwind computer, 1951. (Courtesy of MITRE Corporation.)

Whirlwind worked, and got better as the year progressed. By July, Whirlwind could solve a simple equation and display the answer on an oscilloscope, or display animations from an interactive computer graphic game. Best of all—and a world's first—Whirlwind's computations were done in real time.

Through it all, however, a rug ready to be pulled out at any time from under their well-worked shoes was always at the back of their minds. Forrester penned in his computation notebook that Whirlwind's future seemed less than promising.

December was the cruelest month of all for Forrester, Everett, their band of bright boys, and their machine. The Research and Development Board, the same RDB which now was led by MIT's former president Karl

Compton, had put together in July the Ad Hoc Panel on Electronic Digital Computers. The purpose of the Ad Hoc Panel was to decide which of the 13 government-financed machines should be axed because of limited funds or lack of a compelling mission.

Testimony was sought and heard from experts, investigations were ordered and tours conducted of each computer facility, including Whirlwind. Three months of fact-finding and inquiries culminated at Harvard's *Second Symposium on Large-Scale Digital Calculating Machinery* in September 1949. No experts or character witnesses presented on behalf of Whirlwind.

The Ad Hoc Panel's preliminary report in December eliminated Whirlwind.

According to the preliminary findings and recommendations of the "Report on Electronic Digital Computers by the Consultants to the Chairman of the Research and Development Board," Whirlwind and the effort spent building it were "out of all proportion to the effort being expended on other projects having better specified objectives."

The report's recommendation: cease funding the machine. It was too expensive, and no one had or could think of any use for it.

In a last-ditch effort to sway the panel away from its fatal decision, Forrester and Everett would hit the typewriter keys once more to produce *Report L-24*, titled "Statement of the Status of Project Whirlwind Prepared for the Research and Development Board." It didn't cut any ice. The panel's final report and recommendation for the bright boys' machine in June 1950 would remain unchanged.

Also that December, after the ONR's director Captain J. B. Pearson asked Forrester to project future costs, which Forrester did with a 1951 estimate of $1,150,000, a major kerfuffle erupted. First came the ire of R. J. Bergemann, physical scientist for the ONR's Boston branch office, complaining that the ONR was not getting its money's worth out of Whirlwind, and that Raytheon's Project Hurricane Computer (later known as RAYDAC) was technically superior.

Bergemann was quickly followed by Charles V. L. Smith, who found Forrester's refusal to recognize that there were no available funds "appalling" and found that Whirlwind's program was "excessive."

Mina Rees and Smith propounded Whirlwind's 1951 budget at $250,000 with a little wiggle room to no more than $300,000. Worse yet, the Navy indicated that it was ending its commitment to Whirlwind after fiscal year 1951. Rees would personally journey to Cambridge to officially deliver the news.

According to Kent Redmond and Thomas M. Smith's *Project Whirlwind: A Case History in Contemporary Technology* (1975), entreaties for help from Forrester to Provost Julius Stratton and others in MIT management got no better than lip service.

Memory Breakthrough

Surrounded as he was by the certainty of disfavor from former benefactors and the uncertainty of what next technically, Forrester kept his wits about him and blocked out the unsavory distractions looming ahead for him in the new fiscal year.

To keep a clear head while reflecting over a magazine ad for Deltamax, debating with himself about if and how the alloy could work as a substitute for electrostatic memory, all the while battling day to day upheavals surrounding Project Whirlwind, was remarkable.

He filled his notebook from the spring of 1949 with clear-headed, incisive notes regarding the exact three-dimensional configuration of his new memory concept. His notation of June 13, 1949, displays the exact x, y, z axis control that would eventually become the magnetic-core memory design that would dominate computer memory through to the 1970s.

A revised drawing in his notebook from October 9, 1949, shows a stack of cores that was very nearly his famous final design.

The Nebraska Mr. Fix It and the engineer who was known to his contemporaries as aloof, brilliant, and imperious both came together over that ad for Deltamax. With a single idea Forrester would reinvigorate the whole of computing.

As George Valley remarked years later: "I do not think anyone else would have developed core storage, had Forrester not done it. The people in other organizations who said they were doing it seemed too dilatory. The history of computer memory would probably have been that of transistor storage followed directly after storage tubes."

Innovating as they did such critical new technological advances as real-time computing, parallel processing, reliability, modular machine design, electron tube manufacturing, electrostatic tube construction, and magnetic-core memory—all pulled off in an incredibly compressed period of time—is without parallel in the history of computing. And in 1949, the bright boys were not even half done yet with their string of innovations.

A close look at what was really happening in the Barta Building or a quick glance through their research notes would have quickly

Jay Forrester with magnetic-core memory. (Courtesy of MITRE Corporation.)

differentiated Whirlwind from the IAS machine ... or any other machine anywhere. It would have made the offer for fiscal year 1950 as preposterous to the ONR as it was to the bright boys.

Forrester wasn't the only one who walked around dreaming about magnetic-core memory. Twenty-year-old An Wang (Chinese-born, Harvard research fellow, who later founded Wang Laboratories) had his magnetic-core epiphany, he said, while walking through Harvard Yard.

A recent Harvard PhD in applied physics, Wang started working for Howard Aiken's Mark IV in May 1948 (a year prior to Forrester spotting the Deltamax ad) and was given the job of finding a way "to record and read magnetically stored information without mechanical motion."

Wang's June 29, 1948, notebook entry talks about storing data in donut-shaped magnetic cores, then reading the data from and rewriting data back into them. He found a Navy publication that described the characteristics of Permanorm 5000-Z, liked what he read, and then hunted it down by its U.S. trade name, Deltamax.

Wang's revolutionary idea produced a core memory system for the Mark IV called a magnetic-core delay line (think mercury delay line without the mercury). Putting digital data into a small donut of magnetic material instead of into a huge glass bottle like an electrostatic storage tube was a technological breakthrough with enormous implications. It was the answer to a dependable, cheap, and easy-to-fabricate memory system; it was what enabled the stored-program computer to become a commercial reality.

Wang's problem was that he set out his cores in a series, like a string of Christmas lights, which ultimately proved to be its downfall. Like the anti-quated mercury delay line before it that slowly bubbled up data one slow piece at a time, so too did the magnetic-core delay line.

For real-time needs like those of Whirlwind, a magnetic-core delay line would never do. On September 29, 1949, just before he patented his core memory invention, Wang presented it as a paper to a Harvard computing symposium. In the audience were bright boys Jay Forrester and Bill Papian.

Although Wang was a PhD with some digital circuit experience, he was not an experienced computer builder like Forrester and Papian. Forrester instantly saw the wrong direction Wang had taken.

Forrester, as well as anyone else who was experienced in building computers, knew immediately that arranging the cores in a series was a backwards step. Accessing data would be stable and safe, but would also be way too slow. Fast was the correct direction for the future. Forrester also determined after testing Deltamax that it was not the right magnetic material to use.

He had already bent, kneaded, and shaped Deltamax into rings; he had already experimented running an electric current through the rings; and he had already concluded that the material was too slow at switching current and its physical composition was too sensitive to endure the rigors of mass production.

Forrester laid his cores out in a matrix of 32 cores by 32 cores, and joined them together with interconnecting wires. The matrix design made any core instantly accessible. And it was very fast. The superiority of Forrester's matrix was undeniably self-evident to anyone who saw it.

When Wang learned what Forrester had done, he called it "brilliant" and always regretted not thinking of it himself. An Wang took his core magnet patent and started a company in 1951. Forrester didn't.

With the walls closing in on the Barta Building, it would have been a perfect opportunity for Forrester to dash off with a deep-pocketed investor and a couple of sharp lawyers, and hang a sign out that read something like "Jay's Magical Magnets." It would have been hand-over-fist millions for the young engineer, but he stayed put. He thought only about his mates and his machine.

Instead, MIT would eventually realize $25 million in royalties from magnetic-core memory; the largest royalty haul in the school's history.

With his design set and his proof of concept complete, Forrester next sought out better cores. In 1949, he got Ernst Albers-Schönberg of General Ceramics to brew up hardier magnetic materials. A year later, Albers-Schönberg came up with a magnesium-manganese ferrite material that suited all of Forrester's needs.

The Gang's All Here

Serendipity stalked George Valley during December 1949. Although his thoughts were squarely on radars, there was an ever-enlarging empty spot of concern in his radar plans; a baffling problem that sat in the middle of everything.

The air defense dilemma facing Valley and Marchetti had two parts. First, how to efficiently and very quickly transport a glut of radar information from one or multiple radar sites to a control center. And secondly, once the information got to the control center, how then to reassemble the information into a comprehensive picture of an overall air situation.

Whether for Saville's 75 Lash Up radars, or even for a single radar, the old way, by telephone, was impossibly slow and too laborious to grease pencil telephoned coordinates onto a Plexiglas wall. Utter bedlam would be the result of radar operators shouting coordinates over telephones for wave upon wave of incoming enemy bombers.

Also gnawing at the pair was the vulnerability to low-level attacks coming under the radar screen. Long-range surveillance radars, like the AN/CPS-5 (generally referred to as Microwave Early Warning or MEW radar) had an effective range of 60 miles up to 35,000 feet. In good weather they might scout 200 miles out. However, in order to avoid clutter and false returns from buildings, mountains, and other terrestrial objects, these radars were tilted upward. As such, 5,000 feet of airspace was left exposed to low-level air attack.

The solution of volunteers by the thousands manning Ground Observer Corps towers relaying information from binocular sightings, especially at night, was too frightening to think about. Also, with the near-term likelihood of jet aircraft replacing prop bombers, 60 or even 200 miles of warning would be insufficient. Beginning in December 1949, 24 locations on Saville's list of 75 were turned into temporary Lash Up sites, each with a CPS-5 (distance finding) as well as a TPS-10 height finder (altitude finding) radar.

ADSEC's job would be to come up with a plan to convert all 75 locations into a permanent air defense perimeter; permanent—and certainly more capable than the Lash Up sites. Valley and Marchetti were at an impasse as to exactly how to pull that off.

Part of the answer already existed at Marchetti's laboratory. One of his AFCRL bright boys, John Harrington, had been tinkering with equipment that might help. Harrington's Relay Systems Laboratory in 1948 had relayed by microwave a radar video picture 20 miles from Hanscom Field to the AFCRL in Cambridge and reconstructed it on a display screen called a plan-position indicator or PPI.

Microwave towers in the late 1940s were a distinct rarity, expensive to build and to maintain. Harrington was urged to find another way to relay the radar sightings. Telephone lines were suggested because they "were abundant, available everywhere within the United States, and much less expensive than microwave transmission." The problem was that thin copper telephone lines carried little information compared to microwaves traveling tower to tower through the air.

Harrington theorized that by utilizing the new technology of binary digital transmission, wideband microwave could be replaced with narrowband telephone lines. He would need to convert analog radar data into 1s and 0s, send it over noisy telephone lines, and then reconvert the data back into an analog signal for a CRT (cathode ray tube) display at the AFCRL. His efforts paid off with the development of digital radar relay or DRR, which in late 1949 he tested by sending radar information from MEW radar at Hanscom to a display scope in the AFCRL.

DRR still needed work, but Harrington and his fellow bright boys had pulled off a world's first by automatically detecting radar targets, encoding target coordinates, storing them, and then transmitting them—albeit at a low rate—over telephone lines to a remote site for display or for further processing.

Having seen Harrington's DRR system at work, Valley, one evening late in the same year, hit upon a solution for low-level, under 5,000-foot detection. Doodling while correcting student papers, he scribbled out a triangulation pattern for three continuous wave (CW) radars to detect the position of low-flying aircraft.

CW radars, as their name implies, continuously emit a microwave beam. They were small, like the size of the magnetron wave generator in a microwave oven.

"I quickly discovered that three CW radars, emplaced at different known points [atop telephone poles], could provide signals from which the position of a target seen by all of them could be computed." By "compute," Valley was thinking of a human operator leaning over a display screen with a book of logarithms and slide rule in hand. Marchetti loved the idea, but both of the men knew that the amount of radar data continuously bombarding a human being from CW radars would be overwhelming. "Well, I added lamely," recalled Valley, "maybe we could feed the output of the scaling circuits into a digital computer, and ... "

He was cut short by Marchetti, "Now you're talking, George!" The thought of letting a computer do all the counting was like a revelation to Marchetti.

With a digital computer, they now felt free to CW as many telephone poles as necessary. That was a Saturday in December 1949, and by noon, they had worked out a diagram with CW radars spaced every 10 miles.

They next began the hunt for a digital computer to do all the counting, which would prove to be a very disappointing adventure.

"We did some phoning the following week," to companies, universities, and others known to have built a computer, or to be in the process of building one. "The name Whirlwind was not mentioned by any of my informants," and "their replies were uniformly discouraging: too much time and too much palaver, not to mention the astronomical prices they quoted."

Valley was struck by the fact that hardly anyone thought about connecting a computer to the real world. Most used their machine to perform mathematical problem solving or to manipulate lists. Further, he found none had addressed the need of converting analog signals into digital data. With an ADSEC meeting scheduled for January 20, 1950, Valley and Marchetti wanted to bring to the party a digital computer ready and available to connect to Harrington's DRR system. They resigned themselves to attend that meeting empty-handed.

It would have been a grand opening to the next ADSEC meeting, and a wonderful announcement to herald the upcoming decade of the 1950s, if they could introduce a digital computer to the ADSEC mission.

Just the sound of the words, "the 1950s," had a fresh, newly minted ring. And 1950 was the decade's first marvelous strut on stage. The vision of that first step seemed somehow to shake off much of the dark patina of gloom and warfare from the previous decade. Anticipation for it seemed to be building throughout 1949.

From coast to coast the country seemed primed for something big to happen. The aspect of a new decade prompted a sense of urgency about getting on with life now that the war's end was four years back. Certainly enough time for the war weary to have shaken off the fact that they had survived it all; enough time to have gone the GI Bill route, selected a career or a means of earning a livelihood, found a mate, a nice home in a nice neighborhood for 2.4 people, begun a baby boom, and carved out weekends for coffee over the funny pages and Sunday rides in the family Plymouth.

The United States was like a rocket—engine ignited, straining in its gantry, and ready for liftoff. Europe, on the other hand, was looking at another miserable decade. England had just stopped rationing clothes, bulldozers were still clearing rubble from cities on the Continent, and the Red Army was still threatening every border.

The *Third Man's* Harry Lime was skulking the ruins of Vienna selling bad penicillin on the black market, and George Orwell's Winston Smith got a job from Big Brother rearranging history at the Ministry of Truth in *1984*. Poor old Europe, squat in the middle of the killing fields for six years, was devastated, staggered near mortally, and struggling with all its might to coax a thin flame of hope into illuminating an entire continent. Europe was waiting for a miracle to happen; America was making one.

America was Broadway's *South Pacific*, splashing itself over the country with excitement and verve as Mary Martin sang about washing "that man" and everything else right out of her hair. America was agog with the exotic splendor and vivacity of faraway places with funny names from where GIs had just returned.

Adapted from James Michener's best-selling, Pulitzer Prize winner, *Tales of the South Pacific*, the play like the novel was set in World War II Polynesia, and adapted to the stage by Rodgers and Hammerstein. Broadway transformed the grisly experience of war in the Pacific into musical theatre of palm trees, warm breezes, luaus, romance, and the crooning of Ezio Pinza.

Europe's uncertainty about itself dripped from the eloquent fatalism of Harry Lime's famous lines: "In Italy for 30 years under the Borgias they had warfare, terror, murder, bloodshed—but they produced Michelangelo, Leonardo da Vinci, and the Renaissance. In Switzerland they had brotherly love, 500 years of democracy and what did that produce—the cuckoo clock."

America was a freshly laundered Hawaiian shirt with nothing standing in its way. And America would have over three decades more with which to saunter through it all before the rest of the world caught up. No one figured on the North Koreans coming along to put a few major wrinkles into that crisp shirt.

America's break into the open was led by technology, and its tremendous potential to create wealth. Temptingly, it beckoned to those who recognized the opportunity and seized upon it. Thereafter, fabulous technological innovations begat fabulous fortunes that begat the mega-giants of corporate America.

Electronics led the pack, and there was no better place to witness the parade going by than from the sidewalk in front of the Barta Building.

Word of the future king, the transistor, was on the street. When announced at Bell Labs' headquarters the previous June, it made hardly a ripple of news, winding up buried way inside the *New York Times*. That the thumbnail size device could substitute for an electron tube, require no hot filament, and last nearly forever was a big event for the bright boys. It filled a huge hole in the future of computing by validating that the shakiest part of an electronic digital computer, its banks of unreliable glowing tubes, had a solution that was only a matter of time in coming to pass.

First used in hearing aids and transistor radios, the transistor would take a few years before wending its way into a computer. A 1949 MIT graduate student, Robert Noyce, would take considerable notice. Noyce would later gang transistors together to co-invent the integrated circuit, the forerunner of modern microprocessors—the stuff that put the boom into Silicon Valley.

Information theory rumbled down Massachusetts Avenue as Claude Shannon's "A Mathematical Theory of Communication" showed that "something as seemingly abstract as information could be measured and quantified—that it is intimately linked to thermodynamics, that Nature seems to speak in the language of information," that radar signals, voices over a telephone, and the data coursing through Whirlwind's heart were all the same.

When Valley witnessed Harrington's work with the digital transmission of 1s and 0s as radar pictures over voice telephone lines, he remarked that it "was the first practical demonstration of information theory that I had seen." Norbert Wiener's *Cybernetics* showed the relationship between information and control, and that the information in what he called "feedback"—whether in humans, animals, or machines—contained "messages" used for control. His cybernaut, the Greek word meaning "steersman," uses the message information from feedback to "steer" "through the random forces of the physical world, based on information about the past and forecasts about the future." The works of both Shannon and Wiener are marvelously brought together by Howard Rheingold in his *Tools for Thought*.

"Common words like communication and message were given new, technical meanings by Wiener and Claude Shannon, who independently and roughly simultaneously demonstrated that everything from the random motions of subatomic particles to the behavior of electrical switching networks and the intelligibility of human speech is related in a way that can be expressed through certain basic mathematical equations."

What wonderful New Year's presents for the bright boys were these lines of convergence. The boys had grabbed onto the right star and were on the ride of their lives. Their work was as new as new can be and astoundingly important to boot.

They were at the heart of something really big, manipulating Shannon and Wiener's work on a daily basis. All the pieces of the digital future were tumbling together around them. These bright boys were defining a new industry and giving themselves new careers within it. As with most new things, not many others could see it as well—and for many more not at all.

The bright boys had the digital eyes. Everyone else was on the same sidewalk but sporting 20/20 vision and seeing only traffic jams and pedestrians.

The New Year put no bounce into George Valley's step as he shuffled through the halls of MIT late in January 1950. He and Marchetti were still without a computer. It was discouraging.

Bumping into old friend Jerome Wiesner in the hallway was just the elixir he needed to get revved up again. Wiesner, at 35 years old, was an electrical engineer in the Research Laboratory of Electronics. He was a Rad Lab grad, an expert in microwaves and radar, a decade distant from becoming science advisor to President Kennedy, and a good couple of decades distant from becoming MIT's thirteenth president.

Valley's run-in with Jerry, as he was called by his buddies, was most fortuitous; he could not have planned it any better. Wiesner was brilliant, compassionate, saw the big picture in things, was no slave to accepted opinion, and loved new ideas and breakaway thinking. George's tale of radars, computers, and air defense got Wiesner's motor running.

"I told him about connecting radars and computers, and that I could get money to make a test if I could find a computer whose proprietors weren't too crazy or too busy. He immediately replied that one was up for grabs, right there on the MIT campus." Wiesner then went on to add the name Forrester to the name of a machine called Whirlwind and that both haunted a place called the Barta Building, which Valley passed often because it "was halfway between my office and Marchetti's."

Wiesner didn't tell him much else, nothing about Forrester's problems or the machine's downside. Wiesner did one better; he arranged a lunch where Valley and Forrester could meet face-to-face and figure things out for themselves.

Several days thereafter, on the Friday afternoon of January 27, 1950, Valley and Forrester probed one another over warm dinner rolls, while Jerry Wiesner, his matchmaking completed, faded quietly back to the halls of MIT.

What Valley was up to did not come as a surprise to Forrester. Seven days earlier, Perry Crawford, while visiting Whirlwind, told him that ADSEC had recently gotten underway. ADSEC, he added, was very quickly taking on a new, unofficial moniker as the Valley Committee. Forrester also had a pretty good idea as to its mission for the Air Force.

Valley had done some sleuthing as well. He had checked around about Forrester and his machine: "All these people gave us reports that differed only in their degree of negativity." Marchetti called the ONR to get more information, a call that hinted that the ONR was ready to drop the machine entirely. Marchetti and Valley were undaunted by the stories circulating about the Barta Building as a house of horrors.

Valley plainly and completely laid out for Forrester all the specifics. Valley's need for Whirlwind must have done wonders for Forrester's spirits, not to mention his blood pressure. What a New Year's gift for a harried heart.

It wasn't he and Bob Everett beating every bush from Cambridge to Washington to find a willing sponsor for their machine; it was refreshingly quite the opposite: two men breaking bread together on a wintry Friday afternoon, honestly trying to mesh their talent with their hopes in order to find success together.

Forrester wrote later in his notebook that Valley seemed truly interested, and his honesty and frankness were refreshing as well.

Not wanting to spook Valley's interest on a first date, Forrester tip-toed around the ever-present issue of Whirlwind's fragile memory. Everything else about the machine seemed to be winging along except for that one trouble spot.

Electrostatic memory tubes had been the machine's bane since Forrester's first investigations with the Panel on Electron Tubes back in 1946. And here he was four years later still fumbling at a solution.

He had checked on Raytheon's storage tube project, Edgerton's glow discharge tube, and the famous Williams-Kilburn tube, after which, by 1948, he was forced to make his own.

Of the first 20 electrostatic storage tubes made by the bright boys, only 13 passed muster. A current cost analysis of tubes with storage capacity of 256 bits each was projected to run as high as $1,500 per tube at a production rate of one and a half tubes per week.

Although he was writing in his journal about magnetic-core memory throughout 1949, the most reliable memory around was the antiquated but workable flip-flop memory. Flip-flops were as old as the Eccles-Jordan circuit from 1919. Flip-flop memory was basically two electron tubes that could flip-flop their states from binary 0 to 1 and back again.

In terms of storage, a flip-flop was a 1-bit memory device. It would take 256 flip-flops to equal one of Forrester's electrostatic tubes. That was a horrid thought.

Worse, if the current was turned off, all memory was lost as well. Flip-flops as switches for Whirlwind were passable devices, but as memory they were worse than a finger in a dike.

By March of 1950, the bright boys had 16 of their own hand-built, electrostatic storage tubes in Whirlwind. For a group that prided itself on totally reliable components, these memory tubes were an exasperating reality. Whirlwind as well as every other electronic digital computer had somehow to ditch glass bottles forever.

At lunch with George Valley was the big break that Forrester and his bright boys had long sought. He now had to call out the inventor in himself and quickly get about pulling the rabbit of magnetic-core memory out of his hat.

Forrester invited Valley back to the Barta Building.

Valley was overwhelmed by the size of Whirlwind; he had never seen that many electron tubes in his life. And the machine "was functioning: it was calculating a freshman mechanics problem and displaying the solution on a cathode-ray tube," wrote Valley. He was hooked.

Forrester and Everett loaded him down with the *L1* and *L2* reports, quarterly reports on the Watson Laboratories air traffic control work, memoranda by Gordon Welchman, and David Israel's proposal for his master's thesis on air traffic control. Valley made an appointment to return the following Monday with a few of his friends and associates from ADSEC.

Interior of the Barta Building at the Whirlwind control room, circa 1950: (standing left) Jay Forrester and Bob Everett; (seated) Stephen Dodd; (far right) Ramona Ferenz. (Courtesy of MITRE Corporation.)

Marchetti wanted to know all about the visit. Was the machine as bad as everyone had told them it was? "I replied that while Whirlwind gave the appearance of being mechanically overdesigned, and also looked like something guaranteed to set the teeth of experienced pulse-circuit designers on edge, it nevertheless seemed to work, and it was available." Valley spent all of Sunday going over the tomes that he lugged out of the Barta Building. He was impressed with the minds who had authored them.

On Monday, January 30, 1950, Valley returned to the Barta Building, accompanied by Marchetti, and ADSEC members Eugene Grant, Charles Draper, and Guy Stever.

Forrester and Everett got Whirlwind to perform. They all nodded, shook hands—and the air defense of North America began.

Chapter Six
Island in the Stream

The gang was all there. In a most improbable place, a most improbable convergence would come to pass. Who could have reckoned that all the right pieces from all the right people would each find its own way to the drab old building on Massachusetts Avenue at exactly the right time? Who among them could foresee that what each could never do alone, they would now do together, and in the process become a juggernaut of innovation and change?

Marchetti would roll out his radars. His colleague, John Harrington, would then hook them together with his digital radar relay (DRR), converting digital signals into analog and then back into digital, sending them over telephone lines from Hanscom Field to Cambridge. Whirlwind could then gorge itself on the digital feast, spitting back aircraft position information onto radarscopes.

Saville and Fairchild would keep the Air Force in play and the checkbook open. And Valley, as technology impresario, would manage everything steadily up the totem pole of awareness and acceptance. Their air defense system would blossom into the world's first great electronic system and come to be known as SAGE, for Semi-Automatic Ground Environment.

It would be an electronic system like none other before it: a veritable Book of Genesis for all systems builders to come, just like its famous forebearer—Whirlwind. A chip off the old block, but a giant one!

With Whirlwind the boys wrought the first digital network; the first manifest display of information theory; and

Cambridge Field Station (later, Air Force Cambridge Research Laboratory) 224 and 230 Albany Street, Cambridge, Massachusetts, 1948 (around the corner from the Barta Building). (Courtesy of the United States Air Force, Hanscom AFB.)

the first coming of the now familiar tools of Information Technology: real-time digital computer, keyboard, light gun (early mouse), monitor screens, printer, software programs, real-time memory, modems, and network connections.

With SAGE, the boys would bust out of the Barta Building to build the first information empire. And with it the age of high-tech warfare would begin.

When yoked to its giant peripherals—microwave radars from the bright boys at the AFCRL—the network-within-a-building that was Whirlwind would plump out into a network-within-a-region, and later, swell into a network-within-a-continent. By any measure, it was a mind-boggling ex-pansion, especially given the fragile state of digital electronics in the early 1950s.

From a single Whirlwind behemoth, 48 advanced replicas of the 55,000-vacuum tube, 32-bit computer were cloned (redesignated in military nomenclature as the AN/FSQ-7). Two each were housed in 23 four-story concrete bastions (plus one more duplex set some 700 feet

Interior of the SAGE Combat Center at Hancock Field, New York. (Courtesy of the United States Air Force.)

underground at North Bay, Canada), each 150′ by 150′ square with 6′ thick walls and covering 130,000 feet of floor space, each with its own equally gigantic power and cooling systems, each consuming a white-hot three megawatts of electricity, and each sprinkled strategically all over the country; enough to provide air defense coverage for both the United States and Canada.

Linked together by dedicated telephone lines, eerily lit blue rooms of glowing monitor screens tracked everything moving in the sky, fielded calls from 8,000 Ground Observer Corps outposts, and monitored search radars aboard Navy picket ships, on airborne Warning Star radar aircraft, and on oil rig–type ocean platforms called Texas Towers.

Across Canada, three continent-wide belts of radar installations at 49, 55, and 70 degrees north latitude—the northern most being the famous 63-station Distant Early Warning (DEW) Line from western Alaska eastward across Canada to Greenland—fed radar data into the SAGE system.

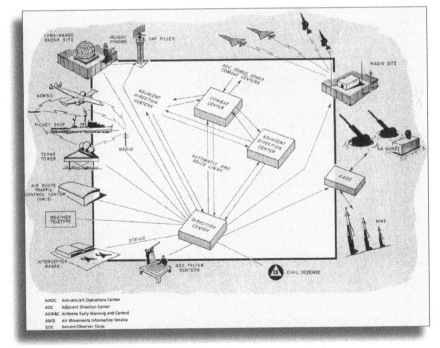

SAGE information network schematic. (Courtesy of MITRE Corporation.)

In all, over 181 manned and unmanned radar sites, each with either AN/FPS-19 L-band search radar or AN/FPS-124 Doppler radars for low-flying targets, chirped out aircraft sightings southward into the giant AN/FSQ-7 computers for processing.

Once intruders were spotted, SAGE could also call out interceptor aircraft. The system "was designed to control the interceptor after takeoff, direct it to the target, and bring it back to a final approach to its runway."

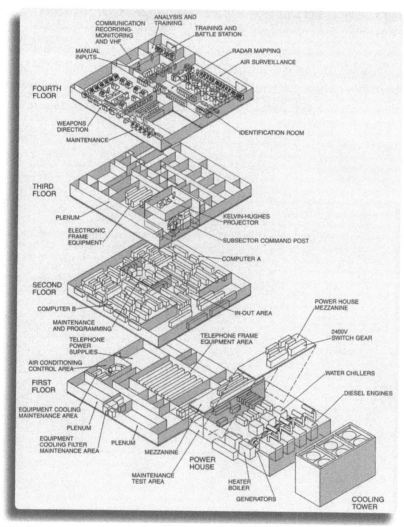

SAGE direction center schematic. (Courtesy of MITRE Corporation.)

All of this round-the-clock tracking, reporting, and interception, however, was expensive: the telephone bill alone for SAGE's private, land-leased lines would by 1959 approach $200 million a year.

SAGE would be an expensive but versatile sky warrior. It also tracked and reported weather conditions, controlled interceptions using NIKE surface-to-air missiles, and later, targeted 500 BOMARC high-altitude, surface-to-air missiles from ten launch sites, which, at $14 billion for the BOMARCs (built by Boeing), was nearly twice that of the entire SAGE system, at $8 billion.

The bright boys even acted as troubleshooters for Boeing's electronics. One critical discovery early on was that the BOMARC's (IM-99A) test-to-operate switch was kluge, that switching a single rocket to erect and fire actually switched all the rockets to erect and fire. Boeing was quick to fix the situation, recalled bright boy Les Earnest, who discovered the switch problem, wryly recounting how he "published an analysis … with the provocative title 'Inadvertent Erection of the IM-99A.'"

In 1954, General Curtis LeMay, commander of the Strategic Air Command (SAC), recognized the implications of SAGE's real-time information flow for command and control. He opted to dump SAC's slow teletype circuits and radio links. A SAGE descendant, the 465L system, was built for the Air Force and tucked away into the bowels of Colorado's Cheyenne Mountain, home today of NORAD, North American Air Defense.

LeMay's enlightenment thrust the Air Force into an upward spiral of ever-improved, real-time information systems, leading to the World Wide Military Command and Control System, the Strategic Automated Command and Control System, and eventually to the early glimmerings of the Internet. In the process, the Air Force was "completely transformed from a group that initially resisted computerization to the leading advocate of computerization within the military."

Indeed, there were unseen worlds within Whirlwind.

Bob Everett, his crew's carefully crafted report *R-127 Whirlwind I Computer Block Diagrams* from 1947, and the bright boys' prudent quality control during Whirlwind's manufacture and management would be poised to explode into a Big Bang of information networking.

Peripherals for the bright boys to connect up would number in the thousands, including far distant radar sites, communications systems, and multiple SAGE blockhouses spread out over millions of square miles of geography. Each device, from the smallest modem and light gun to racks of telephone switching bays to huge search radars—every one of which possessed its own electronic subsystem—would all be knitted together into an immense electronic tapestry, a system of systems the likes of which had heretofore never been contemplated much less ever attempted.

Where once there were but a handful of computer programmers in the entire United States, thousands were hired and trained by the System Development Corporation (SDC)—a SAGE subcontractor—to build SAGE's command and control program, prompting SDC's later claim that it alone trained the country's programming industry.

Light gun and SAGE tracking scope. (Courtesy of MITRE Corporation.)

SAGE was the first system to innovate computer time sharing, where multiple computer consoles could simultaneously share the same machine. When SAGE operators used a light gun (like a modern mouse and cursor) to touch a monitor screen and transfer information into or out of computer memory, it was the first time humans directly manipulated a computer. If the bright boys' dream was to build Whirlwind as a control computer and not just a number crunching machine, then this was it in spades.

SAGE would not be perfect; it would be flawed and vulnerable and highly controversial, yet there it would be in the flesh for all the world to see, a giant loom spinning information nonstop across a continent and seemingly everywhere in between. This was a system to ape for an as yet unnamed Information Technology; a model upon which an industry could arise.

Two years removed from Claude Shannon's famous paper, follow-up work about the true nature of information would just begin to twitch in its chrysalis with the likes of physicist Donald MacKay at the first *London Symposium on Information Theory* in 1950.

A retrospective on business management for *Dun's Review* in 1958, would point out: "Only in the past dozen years [since 1946] has the concept of information—as distinct from the papers, forms, and reports that convey it—really penetrated management's consciousness. That it has done so is largely due to recent breakthroughs in cybernetics, information theory, operations research, and the electronic computer."

The machine side of real-time information processing, however, would still be a chancy business and lag behind Whirlwind. As late as 1961, real-time digital hardware was still being approached with trepidation, as Sylvania engineer W. A. Hosier cautioned in his apprehensive paper, "Pitfalls and Safeguards in Real-Time Digital Systems." Real-time computers, he wrote, "are among the most complex and delicately coordinated of modern engineering enterprises and involve a wide variety of equipment and techniques that frequently tax the state of their respective arts."

He reasoned that if real time wasn't necessary for the computational tasks at hand, then why bother? Most didn't. Whirlwind got lucky; it had found SAGE and found in the military a need for fast information. The world would take some time catching up. However, SAGE would sooner than later spawn totally new industries, open new worlds for scads of manufacturers, and hurtle thousands of individuals into totally new careers. The bright boys created an island in the stream of electronics; an island that would in time become the land.

Civilian authorities similarly recognized an important ability in SAGE: if SAGE could spot every "unfriendly" aircraft over North America, it must then be able to spot and track every "friendly" aircraft as well. Tracking commercial airliners, it was reasoned, might keep them apart and thus avoid midair collisions. Time magazine reported in early 1957 that "in the last four months of 1956… there were 452 near-misses between airborne planes."

It was only a matter of time until a near-miss became a catastrophe. That same year, a midair collision over the Grand Canyon between United Airlines Flight 718 and Trans World Airlines Flight 2 killed all 128 aboard both aircraft.

Enter SAGE, opening the skies to a new way of controlling and managing commercial air traffic, from which modern air traffic control directly descends. The Federal Aviation Administration (FAA) would eventually build its own SAGE-like system to control the national airspace, which, in turn, also became the model for international air traffic control. A SAGE control center in Montana shared its computer and display screens with the FAA for nearly 15 years, serving as the FAA's Great Falls Air Route Traffic Control Center.

IBM, manufacturer of the SAGE AN/FSQ-7 computers, not only pulled in $500 million for building the computers (its biggest single contract of the 1950s), but also built the first electronic digital computer assembly line at IBM's Kingston, New York, plant, and an electronic development laboratory in Poughkeepsie, New York. Between 1955 and 1957, IBM employed over 7,000 workers to build SAGE machines.

IBM also used its SAGE real-time expertise to push out the world's first airline flight-booking system, the Semi-Automated Business Research Environment (SABRE) system, in which Forrester's old pal Perry Crawford was ensconced helping out the IBM effort. Airline reservations that previously took nearly three hours each to book were made in real time as the SABRE system prowled its database to match flights with customers.

The invention of the SAGE database, as well as the coining of the word "database" itself, were two other firsts claimed by the System Development Corporation. "SAGE had to present an up-to-date and consistent representation of the various bombers, fighters and bases to all its users.

"The System Development Corporation (SDC), a RAND Corporation group spun-off to develop the software for SAGE, had adopted the term 'database' to describe the shared collection of data on which all these views were based." Today, of course, every information system has a database, and databases are central to the running of most any business.

SAGE not only qualified SDC's programmers but also trained hundreds from Western Electric at Murphy Army Hospital in Waltham, Massachusetts, to become part of 50-member test teams that went from SAGE blockhouse to blockhouse getting the sites' electronics up and running.

The bottom line to the outpouring of "SAGE effects" upon the new electronics industry seems to have been this: create an exceptional tool and exceptional people will flock to it and use it to create still other exceptional tools. SAGE was just such a tool.

The list of supplicants to the SAGE shrine that became digital converts included not only IBM, SDC, and Western Electric and its parent AT&T, but also Burroughs (later merging with Sperry Corporation to become Unisys), General Electric, Raytheon, Philco (later bought by Ford Motor to become Ford Aerospace), Sylvania, and Bendix (later acquired by Control Data Corporation).

Les Earnest reckons that SAGE "was about 10 years ahead of general-purpose timesharing systems and 20 years ahead of personal computers and workstations."

In the reckless pace that is technological change, that's some kind of longevity. All of which begs the obvious question about why the Navy, the Office of Naval Research, Mina Rees, and MIT were so myopic that they acted to stop the bright boys cold? Their actions were especially strange since they were not callous corporate stockholders hell-bent only on real-time, tangible profits and as such quick to vote out a risky loser. Rather, this group—consisting of the Navy, the Navy's congressionally anointed research laboratory, a vaunted engineering university, and a mathematician tasked with developing computer technology—was supposed to be flexible, open to change, and future looking. In March of 1950, when needed the most, they showed themselves to be otherwise.

Absent George Valley and the United States Air Force, who could have guessed where gumption alone would have carried the bright boys? Years spent breathing life into a new technology, a dream project that comes along very infrequently in researchers' careers, would most likely have been dashed forever, no matter how bright the bright boys were.

Decades later in a speech at Boston's Computer Museum, former bright boy Bob Wieser recollected SAGE as a once-in-a-lifetime experience for any engineer. "We had the engineer's dream," he recalled, "a nationally important problem that was interesting and difficult but not impossible to solve. We were in a day-to-day contest with Mother Nature. The odds were bad, but we always had a chance to win, and we won all the battles that led up to SAGE." Wieser even chimed in with his bright boy assessment of Whirlwind's effect upon the then nascent electronics industry: "We also won the cause for digital computing," he reminded his audience. "If there's anyone who thinks we didn't win, just go to Radio Shack and try to buy an analog computer."

Popular culture got an eyeful of SAGE when Hollywood went gaga over its mammoth hardware, bulging cabinetry, and flashing lights. No fewer than 34 Hollywood movies and TV programs utilized chunks of SAGE as backdrops and props. What was in the public's imagination as to what a computer should look like and how it should act was totally SAGE transplanted to the sets of *Voyage to the Bottom of the Sea*, *Lost in Space*, *The Time Tunnel*, *The Man from U.N.C.L.E.*, *Fantastic Voyage*, *Batman*, and *Independence Day*; even comedies got into the act with *Spaceballs*, *Get Smart*, *Austin Powers*, and Woody Allen's *Sleeper*. That's a span of 40-plus years of SAGE in Hollywood. SAGE's whirring, flashing racks of iron became the obligatory standard for anyone contemplating the design of a backdrop for any sci-fi or even remotely futuristic movie or TV show. What a hoot it was for any bright boy at a movie theatre or in front of a TV set to witness Hollywood's best cavorting with their much-maligned machine: the "big boondoggle" in the Barta Building.

SAGE became a computer hero! No telling how many times it averted disaster for the world or how many millions of lives it saved, all in pursuit of big-screen storytelling. SAGE's name never appeared in the film credits, but it was nevertheless up in lights on the silver screen.

A much weightier venue, the 1996 presidential campaign, also used a SAGE backdrop. The then-famous computer with decades of screen appearances was added to lend some high-tech credibility to the *World News Tonight* stage set. *ABC News* carted in pieces to sit behind newscasters Peter Jennings and David Brinkley as they speculated on the Clinton–Dole returns. No one need venture into a Radio Shack to see the outcome of the digital revolution. All that was necessary was a comfy chair at home in front of the tube with a TV dinner most any night of most any week.

The Big Day

In 1950, "there was only a fraction of a megabyte of random-access memory on planet Earth, and only part of it was working at any given time." For those few souls trying to usher more of it into being, progress was frustrating and haltingly slow.

In March of that year, Mina Rees and her ONR mates would journey to Cambridge and unwittingly jeopardize about 4,000 bits of that precious resource. A practical mathematician and administrator, Rees' main intention in coming to the Barta Building was to wrest back under control what she and the ONR perceived as an extravagant budget.

She had no particular malice toward memory; she gave it little thought, and thought even less about the far-reaching consequences of her actions. Sixteen hot bottles each with 256 bits of fragile memory intended for Whirlwind's racks awaited her arrival. The bright boys needed somehow to stop her.

In large measure the early progress of the electronic digital computer was dependent on the discovery and development of fast, reliable primary memory, what we today call RAM or random-access memory. RAM is the vital, active memory that computers manipulate on the fly while in operation. RAM briefly holds binary text characters, symbols, and numerical values as the computer fetches and executes instructions.

In 1950, electrostatic tube RAM was the fastest technology available. Whirlwind's first set of 16 electrostatic tubes had a paltry total of 4,096 bits of RAM (8 bits equals 1 byte). Eventually, the bright boys would plump out primary memory to two banks of 16 electrostatic tubes for "a total storage capacity of 2,048 16-digit binary numbers," effectively doubling Whirlwind's primary memory.

Each tube would hold a single digit of the 16-digit binary number, and Whirlwind would make a sweep of all 16 digits simultaneously, taking 2 microseconds to add two of the binary numbers or 20 microseconds to multiply two of the binary numbers.

Everett's block diagrams group designed a 16-digit word length (5 instruction bits and 11 address bits) for holding down the cost of the computer. Many mathematicians of the time, especially those from the ONR, considered the word length way too short to be useful, although the first production IBM personal computer (1981) also used a 16-bit word length.

Such infinitesimal amounts of memory—as well as such slow computation speeds—are nearly incomprehensible when compared 70 years later to a laptop computer that would handily toss about a gigabyte of RAM, the equivalent of a billion bytes (exactly: 1,073,741,824 bytes).

Such a remarkable transformation from bits of memory to gigabytes of memory would not begin until 1970 when the Intel Corporation produced the 1103 silicon memory chip, which possessed a single kilobyte (1,024 bits) of RAM. In terms of physical size, it was on the order of comparing a magnum of champagne to a postage stamp.

And better still, there was virtually no heat; there were no electron guns firing beams through huge vacuums at holding plates, no counting the hours until each hot bottle burnt itself to extinction. However, between the bottles and the 1103, between the 1950s and 1970, there was a wasteland of potential alternatives to tube memory.

Serial memory from rotating magnetic drums was too slow, as were acoustic waves flowing slowly—at the speed of sound—through mercury delay lines. The only interim solution was random access magnetic-core memory, which in March of 1950 existed only in the pages of Forrester's notebook. Profound change in the history of computing lay in Forrester's drawings and in his Deltamax tinkerings with magnetic donuts strung over a grid of wires.

Rees' budget visit would threaten more than the ONR's cash flow to the bright boys; it was disaster in the making for electronics and computer technology. In view of the geopolitical events of the day, with the dread of holocaust, invading armies, revolutions, terrorist threats, and rampant fears of a world running amuck, halting any progress going on in the Barta Building was an ill-conceived roadblock to national security.

"Two conferences took place on March 6, 1950, one in the morning and one in the afternoon." Into the Barta Building first thing Monday morning flowed the MIT contingent led by Provost Julius Stratton, with Mina Rees heading up the ONR group.

Neither Forrester nor Valley nor any of the bright boys attended the morning session. It was billed as a "policy meeting," which was MIT's way of softening up the scene for both the entrance of a white knight and amicable surcease for the cash outflows from ONR's wallet. It was all accomplished with a smile over coffee, without any prancing from a white knight or wiseass researchers gloating in their cups.

The ONR was far from blindsided by the meeting. ADSEC's activities, ongoing from January 20, were well known, as were those of the Ad Hoc Committee on Air Defense, as well as Marchetti's inquiries to the ONR about Whirlwind's availability. Each side knew the outcome well before the meeting began; all that was really necessary was a formal face-to-face and some polite, academic-style niceties.

The afternoon session would be ideal for a gloat or two, seeing that Rees' Charles V. L. Smith, who had savagely ripped Forrester's budget projections, was there for the changing of the guard. Forrester and Valley were ready for the event; Valley had Air Force authority to spend, and a $500,000 opening offer to throw on the table.

It was an eye-opener, and very quickly "all agreed that this would be an excellent solution to the situation." Valley said later, "I could have gotten much more, but that's all they seemed to want at the moment." He explained that his needs for Whirlwind were to run experiments whereby ADSEC could test out the feasibility of sending radar signals to a digital computer. The hope was that such an arrangement would give the Air Force the ability to quickly scan, track, and intercept multiple aircraft simultaneously.

It was agreed that both the Navy and MIT would still have access to Whirlwind for their computing needs; the Watson Laboratories' air traffic control contract, including the services of Welchman and his crew, would be transferred to ADSEC; and the ONR's $300,000 for the upcoming fiscal year (1951) would be forthcoming.

Including the remaining budget for the existing fiscal year (1950), Forrester suddenly had continued financing of over $1 million, which in the desperate struggle to keep Whirlwind alive put his survival skills somewhere between those of the enigmatic Harry Lime and the crooning of Ezio Pinza. Rees was more succinct in the reason for the bright boys' close shave: "They were lucky."

The white knight's journey to the March meeting was strangely serendipitous and convoluted, especially for an anti-nuke physics professor, comfortably tenured, easing his way into a stretch of benign course instruction with a little bit of cosmic ray research on the side. In 1950, George Valley was 37 years old with a family and a new home in Belmont.

Gone were his Rad Lab days of flying to London in a bomb bay and winging home in a flying boat by way of Brazil. He had done his bit.

He was not personally challenged at breathing life into a monster computer as Forrester and his bright boys were. He did not need to shoulder the weight of attacking Tu-4s like Fairchild and Saville. He didn't have the all-consuming drive for national security as did Bob Robertson. He could have just waved off Robertson's taxi with, "No time now, I'll call you later."

He could have escaped into the pursuit of postwar normalcy just like millions of others. But there he was in the Barta Building with millions of dollars to spend and carte blanche to spend it. Eyes from everywhere would soon be on him; his reputation would be on the line; he was in deep, right up to his chin, attempting to solve a problem that just might have no solution.

And if the ambitious calling he had volunteered for wasn't enough of an undertaking, naysayers added more. "When word spread … that I had gotten in bed with Whirlwind, a number of busybodies warned me that it was a grave error … and I found myself snubbed in the halls of MIT by a personage very high in its administration."

Valley, as Forrester had done before him, had thrust himself into an awkward situation where, whatever the outcome good or bad, he was there for its duration.

Valley's Mojo

George Valley had major mojo. He was either outrageously fortunate or had the market cornered in lucky rabbits' feet or was the object of divine intervention.

As the leader of a modestly sized air defense project that was about to rocket into the realm of the gargantuan, he never once along the way (from Whirlwind to SAGE, 1950–1958) stumbled into a project-killing event or person, and never fell victim to dead-ends, huge losses in time, lack of money, shoddy engineering, or just bad technology. That's not to say that he wasn't constantly menaced by misfortune, he was—but he somehow adroitly sidestepped every threat. He not only avoided bad fortune but also seems to have had an unerring knack for falling into good fortune.

As luck or rabbits' feet or divine intervention would have it, he experienced a series of near miraculous first encounters with people and events that always proved beneficial and shoved the project forward. Some of it, as Bob Wieser saw it, was due to the quality of people around him and their dogged belief in their ability to pull off the seemingly impossible.

Much also had to do with Valley himself, his inner fortitude, his courage at taking on an ostensibly hopeless job that he could have easily avoided, and the how-to experience he gained from his war years at the Radiation Laboratory. There was something about Valley, something ineffable, something special that always seemed to make things go the right way at near precisely the right time.

Valley's string of first encounters was too remarkable to be mere coincidence. The taxi ride at the Pentagon with Robertson was pure serendipity, but it jumpstarted everything else and set up the encounters that followed. Valley's encounter with the Lash Up system brought him to the doors of the AFCRL's John Marchetti, the Father of American Radar, who (according to Truman's science expert, William Golden) had the best military research facility in the country, loaded with good researchers like John Harrington, who would soon become vital in creating the SAGE network. His quick analysis of Lash Up and his phone call to von Karman were both timely and right on the money.

Through Fairchild and the Scientific Advisory Board (SAB), Valley next encountered the brilliant, handpicked aerodynamicists for his Valley Committee. The Valley Committee meetings intersected him with Gordon Saville (by then a Major General), the country's premier air defense expert. Valley next bumped into Jerry Wiesner in the corridors of MIT, a meeting that produced a very fortunate January 1950 luncheon with Jay Forrester. That bread-breaking encounter was followed up by a visit to a "working" Whirlwind, which got Valley his Air Force money from Fairchild and control of the Barta Building from Rees.

It was a remarkable seven months with nary a hitch, stumble, or false move along the way. And that was only from September 1949 until March 1950; Valley still had an additional seven year of first encounters ahead of him. He would not disappoint.

The Air Defense Engineering Committee (ADSEC) cranked up for business on January 20, 1950, seven days prior to Valley's big intro to Whirlwind at the Barta Building, and a full five months before that watershed moment in preparedness, the outbreak of the Korean War.

January 20 was a Friday; ADSEC would continue meeting on Fridays for the next two years. Very quickly, the ADSEC acronym was dropped in favor of the Valley Committee: it was easier to say and more direct, so it stuck. Besides, the looming figure of George Valley was associated with every aspect of the air defense group and made his name a natural and convenient reference. The Valley Committee met in a rather spartan room at the AFCRL on Albany Street. But it was a handy place to get to for all involved and, better still, was located at the spiritual center of the newly emergent digital electronics.

As a practical matter, the ADSEC site was also selected so as to be near a large city, "so the final product could be used to defend that city as well as to serve as a model for the other installations."

Valley further suggested that his fellow committee members be drawn from the fields of physics, electronics, aerodynamics, and guided missiles. ADSEC had nine permanent members, four of whom were SAB affiliated and also taught at MIT—Guy Stever and Charles Draper were aeronautical engineers with expertise in guided missiles; Henry Houghton was head of meteorology; and William Hawthorn was a mechanical engineer and renowned expert on jet engines. Aerodynamicist Allan Donovan flew his own plane in every Friday from the Cornell Aeronautical Laboratory. Physicist and wartime blind-landing researcher, George Comstock, vice president of Airborne Instruments Laboratory, rounded out the group together with Valley, Marchetti, and Air Force liaison Major Richard Cella. Any week could see an ADSEC meeting swell with any number of invited guests or a specialist lecturer or two.

Valley and Marchetti set out an intensive weekly work schedule for the committee. It began innocently enough with quick reviews of the air defenses of the Royal Air Force and the Luftwaffe and briefings from the United States Air Force on the current U.S. capability. Most of the introductory topics were casual, get acquainted-type icebreakers for the nine-member group.

Subsequent topics were more to the point: a Raytheon ground radar proposal, a Douglas Aircraft proposal for a supersonic jet interceptor, new work in X- and L-band radar, airborne radar, airborne analog computers, a visit to Air Force Headquarters for a Gordon Saville air defense briefing, an investigation of the potential of unguided rockets, the use of telephone lines to forward radar information, and the use of a computer to gather, store, and distribute radar position data.

For the telephone–computer connection, the Valley Committee meeting of February 17 started with Marchetti briefing on the use of continuous-wave radars tied to a computer, which was followed by Jay Forrester pitching Whirlwind and its "10,000 arithmetic computations per second." It was easy for the members to see the simple logic in the use of a computer. "There was no conceivable way in which human radar operators could be employed to make these calculations for hundreds of aircraft as detected from such a large number of radars."

The feasibility of using telephone lines to send position data to a computer was later presented by Western Electric's Claire (Hap) Halligan (future first president of the MITRE Corporation). Western Electric's expertise was based on its astounding 1942 installation of the Pentagon's telephone service (the Pentagon, the largest office building in the world, was built in 16 months), and was the communications nerve center used to direct the entire defense effort for all of World War II. It had the world's largest private branch exchange (PBX) with 13,000 lines of dial PBX equipment, 125 operator positions, and 100,000 miles of telephone cable.

Wiring a continent full of radars to computers was a job that Western Electric seemed ideally suited to take on. Very early in the life of the Valley Committee, Valley and Marchetti were hammering home their conviction for radar signals over telephone lines to an electronic digital computer. It found a receptive audience with the Valley Committee. They voted to include a line item in the budget to rent Whirlwind.

Whirlwind's salvation was assured by February, and Forrester, along with many of his fellow bright boys, became regular attendees at Valley Committee meetings. It was a strangely new presence for otherwise traditional air defense meetings between radar engineers and aerodynamicists; electronic digital computer guys were now in the mix bringing along their own argot of clock speeds, central processors, memory banks, programming, and architectures.

Strange as the initial intrusion may have seemed, there was, however, enough of a commonality among the disciplines to see the inescapable logic and inevitability of engineering it all together. The Valley Committee was for the first time contemplating the building of an intricate, real-time, electronic information system. This was not a railroad system, a waterway system, an electric power system, or even the telephone. This was to be the conscious and systematic manipulation of electrons to pass digital information among and through a series of complex electronic machines in order to form a single, dynamic system.

This was something new. These men were not thinking in terms of radar only or computers only or data transmission only; they were thinking about an entire interconnected system: systems thinking. Systems thinking preceded the actual designing of the interconnected dynamic system: systems engineering.

Simon Ramo, an early systems engineering advocate and chief engineer of the Air Force's Atlas missile program from the early 1950s, best explained it: the function of systems engineering is to "integrate the specialized separate pieces of a complex of apparatus and people—the system—into a harmonious ensemble that optimally achieves the desired end."

Marchetti went so far as to build a scale model of the proposed radar-to-computer system; and the committee voted for June of 1950 "to make an early test by hooking up one of Marchetti's radars at Bedford to Whirlwind, using Harrington's phone-line apparatus."

"At the time, Whirlwind had no electrostatic storage," recalled Bob Wieser, who, along with a group of graduate students that included David Israel, Robert Walquist, Jack Arnow, and Howard Kirschner, switched from working on air traffic control with Whirlwind to air defense. "Random access memory was five flip-flop registers and 32 toggle switch registers that could be read by the machine. We got the radar data into the machine and displayed."

The radar echoes sent from Hanscom Field's MEW Hill radar to Whirlwind in the Barta Building would prove to be not only an important test for air defense but also a singular moment in the birth of Information Technology: information theory, cybernetics, and digital computing all converged with the first packets of digital information sent from Bedford radar over telephone lines to Cambridge.

It was the first step in a digital journey that would culminate when Whirlwind computer descendants, the TX-2 and IBM's AN/FSQ-32 (based on AN/FSQ-7 and AN/FSQ-8 technology), were networked together by Larry Roberts on the East Coast and Tom Marill at SDC on the West Coast. Just beyond that piece of networking lay the shores of the Internet. In fact, Roberts is often called the Father of ARPANET, which was the predecessor of the Internet. It would take more than a decade of work to reach the watershed event of transcontinental computer-to-computer networking. The Bedford-to-Whirlwind detect and intercept experiment in April 1951 came first and would be scaled up to a larger experiment called the Cape Cod System in 1953, which scaled up even larger, becoming SAGE in 1956, and then to the first mammoth SAGE installation at McGuire AFB in 1958.

Forrester and Everett had theorized on such system's ideas in their *L-1* and *L-2* laboratory reports to the Navy in 1947. Louis Ridenour (SAB member and USAF chief scientist) hit close to the mark, authoring in 1947 one of the famous Radiation Laboratory textbooks, *Radar System Engineering*. A case for early concepts on systems engineering can even be made for the original thinking found in the master's thesis of Perry Crawford (1942).

However, nothing previous to the Valley Committee approached the scope of anything like SAGE systems thinking. It is interesting to note that the keen-eyed word hounds at the *Oxford English Dictionary* took serious note by including terms like "systems engineering," "systems analysis," and "general systems theory" in OED editions only after 1950. Later, in 1957, Harry Goode, along with co-author Robert Machol, would be credited with authorship for the early bible on systems engineering, *Systems Engineering: An Introduction to the Design of Large-Scale Systems*.

The Friday, March 17 Valley Committee meeting returned to the nitty-gritty of radar, specifically, "the serious limitations of the new generation of ground radars." Over-water ducting of radio waves, situation displays, and data handling problems were also covered. Forrester attended, weekly getting more familiar with the overall system in which his machine would play the central role. The very next week, March 24, Forrester returned with five of his fellow bright boys; other visiting specialists also joined the meeting as guests.

Forrester discussed the "problem of combining data from three or more radars, and mentioned the need for storage tubes; Harrington discussed terminal equipment for pulse radars that were to be used with a computer." Systems engineering the air defense pieces together into a single, dynamic entity was beginning to pick up speed.

In the spring of 1950, to intensify the hunt for low-altitude radar coverage, Valley began holding additional "Thursday-evening radar seminars to which mathematicians, engineers, and physicists from industry as well as universities and Air Force installations were invited." Picking up speed as well were the ranks of the attendees—the spartan room at the AFCRL was starting to swell with curiosity, interest, and popularity. Valley wrote that ADSEC's "renown began to grow" and that in July he was invited to a Pentagon briefing along with 21 generals and the Secretary of the Air Force.

Barely five months after his Barta Building meeting with Mina Rees, Valley experienced a first encounter with the totally unexpected: fame.

The Competition

The landscape for electronic digital computing in the early 1950s seemed a scary place in which to venture either one's money or career. There really was not much of it going on anywhere, except for a dozen projects that the military had ordered and a sliver of emerging interest from private industry. It wasn't taught in college and mothers weren't urging their sons to take up the cause of the *digerati*.

In actuality, the early 1950s would prove to be the best of times in which to take a chance on a digital future. The bright boys were quite convinced of it. They had endured and come through; others could as well. They felt sure that theirs was a technology that had nowhere to go but up. "Digital computation is a new and growing field and a young and rapidly developing profession," they wrote in their laboratory's brochure. They were sky high over their own newfound calling's sudden success, which was made even sweeter as Whirlwind became quite the tourist attraction, generating in a single year (1953) 990 curiosity seekers. Some 95 industrial companies sent representatives, as did 40 government organizations, 35 academic institutions, and 23 countries.

It was a far cry from the accusingly suspicious faces that had stalked them from 1948 to 1950. The bright boys were sure about the road to success, insisting, "It is not the machines but the people that are the top asset of a laboratory." And, they felt that a digital apprenticeship might well be the key: "Digital computation is a new field which has been successfully entered by men of good ability having no prior experience in the work," they stressed.

Roger Sisson was a prime example of the allure and transformative effect of the Whirlwind experience. In 1948, this 22-year-old kid from Brookline, Massachusetts, with a fresh degree in electrical engineering from MIT was inspired by a Jay Forrester talk on computers, and was hooked. That September, he found himself in Room 212 of the Barta Building sharing space with other bright boys, Charles Corderman, Norman Zimbel, and Harrison Rowe, just across the hallway from the great one, Forrester himself.

He worked on cathode ray tube displays, and then together with Alfred Susskind, another kid from downstairs in Room 124, did his master's thesis: one wrote about analog to digital conversion, the other about digital to analog. By 1950, the new Whirlwind convert found himself on the West Coast in a doctoral program at UCLA and employed at the recently formed Electronic Engineering Company of California, EECO.

Sisson was but one of the scores of brilliant, youthful acolytes drawn to the humming of the great machine and to mingle with its band of brothers at 211 Massachusetts Avenue. Many, just like Sisson, would issue forth from that seminal training experience to form the vanguard of a new breed of computer engineer. Later, SAGE would out do Whirlwind by making digital converts by the hundreds.

However, patrons or benefactors were necessary in order to survive until the field quickened. Electronic digital computing had three types of early patrons: the military, universities, and industry. Saul Rosen's 1969 retrospective "Electronic Computers: A Historical Survey" traces the impact or lack thereof of all three.

Rosen's perspective is a uniquely balanced one. He was a mathematician with an engineer's sense of machinery and circuits, spent the war in the Army's Signal Corps, was a computer builder, a founder of the Purdue University computing laboratory and computer engineering program, and worked for years within the early electronics industry as it strove to push machines out into the marketplace to make a profit. Rosen had lived it all.

With a PhD in mathematics from the University of Pennsylvania, Rosen worked the teaching circuit from the University of Delaware to Drexel to the University of California to Wayne State to Purdue. He labored for industry giants Burroughs, ElectroData Corporation, and Philco. With Philco he was the chief engineer on the world's first commercial, transistorized computer, the Transac S-2000. His famous 1967 book, *Programming Systems and Languages*, was a rite of passage in computing education. And for many years Rosen was an editor of the *Annals of the History of Computing*. Rosen wrote as an eyewitness to the origins of computing, seemingly to every moment of its unfolding.

A concise 23 pages, "Electronic Computers: A Historical Survey" is an insightful insider's diary on the progress of electronic computing baby stepping its way from 1946 to 1966. Remarkable in Rosen's comparison of the technology's machines, their makers, and patrons is how far ahead the bright boys and Whirlwind were from everyone else outside the doors of the Barta Building.

Rosen does not deliberately exalt them as being kings of the *digerati*, but they nevertheless rise to that level as he describes the computing landscape around them.

Rosen begins with the digital font, ENIAC, at its digital epicenter, Philadelphia. In 1950, Philadelphia was trying hard to become America's first Silicon Valley without the silicon.

The great ENIAC was purring away for the Army at the Aberdeen Proving Grounds while its younger brother EDVAC was still coming to life at the University of Pennsylvania. Not far away, and also pushing into existence, was the soon-to-be-famous UNIVAC at the Mauchly-Eckert works, where it was heading toward a rollout date in March 1951.

Preceding their UNIVAC by two years was the Mauchly-Eckert BINAC, the first operational stored program computer in the United States. While BINAC, a mercury delay line machine, worked well in factory tests, it did not ship well and never worked for the customer, Northrop Aircraft, 3,000 miles away in Hawthorn, California. "The poor engineers were constantly working to get BINAC to run," recalled Roger Mills, eyewitness to the struggle. "One side would be running while they worked on the other side. The two sides never worked together as long as I was there."

The BINAC's avowed purpose in life was to be part of the guidance system for the top-secret Snark missile; its enormous size (equal to a one-car garage) precluded its riding copilot for the missile. However, its computing capabilities opened more than a few eyes very wide at Northrop. "The most successful run we made was on a deicing problem for an airplane. Two operators on electric calculators worked for 6 months computing steps in resolving the differential equations. The BINAC did these steps and completed the calculations in 15 minutes." Fifteen minutes wasn't real-time computing but in comparison to six months, it was lightning-like and very pleasing on the fingers.

Early on, Mauchly and Eckert not only produced leading-edge technology but also, and thankfully so for them, a financial backer to cover their inept cost accounting that allowed UNIVACs to be woefully under priced at $250,000 apiece together with promises for delivery that were utterly impossible to fulfill. Harry Straus, founder of the American Totalisator Company, which made the "Tote Boards" that are fixtures at nearly every thoroughbred race track in the country, invested to the tune of $500,000, becoming chairman of Mauchly and Eckert's company.

Computers were a natural add-on to his electromechanical systems for calculating odds, dispensing tickets, and displaying payouts on horse races. Unfortunately, Straus died in an airplane crash in 1949, and American Totalisator's board quickly abandoned Mauchly and Eckert. With loans coming due in January 1950, they hustled their company around the city looking for a buyer: Burroughs and Philco passed; Remington Rand had interest. Hats in hand, Mauchly and Eckert also met with IBM's Tom Watson Sr. to try to work out a future. Watson, who would have nothing to do with anything that did not relate to the punch card business, declined their offer. Watson rejected computer technology, which was more than enough to give IBM leadership in electronic computing for years to come.

But it was not the first time that his neophobia cost his company a leg up in technology. He had also flatly refused patent attorney Chester Carlson's 1947 offer on an electrophotographic printing device that Carlson later sold to the Haloid Photographic Company, which, in turn, changed its name to Haloid Xerox, and still later to the Xerox Corporation, as the enormous success of xerography (photocopying) began to take hold.

Mauchly and Eckert next hightailed it to Florida for a meeting on Jim Rand's yacht, where the owner of Remington Rand—another Philadelphia powerhouse—bought out their floundering company and the rights to UNIVAC, America's first commercial computer, on February 1, 1950. For under $1 million, Rand landed the best data processing computer in the country, together with $1.2 million in existing UNIVAC contracts from the likes of the Prudential Insurance Company, AC Nielsen, and the United States Census Bureau.

The very next year—1951—exhorted along by Watson's son, Tom Jr., IBM would begin a crash course to produce its own electronic digital computer, playing catchup all the while with a machine that it had let slip away. It is rumored that Watson Sr. nearly blew a gasket when Walter Cronkite, anchoring CBS News, televised the results of UNIVAC predicting the Eisenhower upset in the 1952 presidential elections.

The emerging significance of that event was lost on even a news pro like Cronkite. He recalled later that, "I don't think that any of us saw the long shadow in the newsroom at all." Watson had. And he accorded even greater notice to big IBM customer General Electric buying a UNIVAC to do its payroll. Remington Rand would strike Big Blue again in December 1951 when it bought Engineering Research Associate's (ERA) Atlas computer.

IBM's Werner Buchholz remembers it as a time of drop everything and full steam ahead to push out a UNIVAC competitor: "On the 701 we started with the so called von Neumann machine, the IAS machine at Princeton, as a starting point, and of course Whirlwind."

Watson also had in his stable of engineers Nat Rochester, the guy who built Whirlwind's arithmetic unit and who had taken Bob Everett's course on electronic digital computing.

Fortunately for IBM, the ghastly situation was more than salved in September 1952 when it encountered Project High, which was the movement to convert Whirlwind into the AN/FSQ-7. Project High got its name from a necktie factory on High Street in Poughkeepsie where IBMers and the bright boys met for discussions.

IBM's management understood well the importance of getting the (Whirlwind) AN/FSQ-7 deal: "IBM will be recognized as the undisputed leader in the large-scale, high speed, general purpose digital computer field. If a competitor were performing on this contract, that competitor might gain enough advantage to force IBM into a relatively secondary position."

By April 1953, Watson's long-awaited response materialized: IBM rolled out a computer, publicly announcing the arrival of its IBM 701 Electronic Data Processing Machine. IBM was finally in the game. And when it was selected over Remington Rand and Burroughs to produce all of the (SAGE) AN/FSQ-7s, IBM was really in the game.

At the corner of Philadelphia's C and Tioga Streets, the Philco Corporation (which began its business life as the Philadelphia Storage Battery Company) was not overly keen on the electronic computer business, probably because it was the largest and most well-known radio manufacturer in the world. Philco radios were quite literally in every corner of the globe.

What Philco was real keen about was building the first transistor radio. Failing that, they settled instead for building the world's first transistorized computer.

Invented in 1948, "the transistor was expected to become the key to revolutionary advances in computer technology." It wasn't and would not happen for another six years. "Switching speeds were relatively slow; it was difficult to produce transistors with uniform characteristics."

The transistor was shuffled off to serve the hearing aid and radio business, and with radios it met head-on with the radio king, Philco. Getting bulky tubes and heavy batteries out of portable radios was a major effort with radio makers of the 1950s, and to be the first to put out a transistor radio would be a tremendous market coup.

Philco did not produce the first transistor radio: the Regency TR-1 was the first, and that was not until 1955. Worse still, the Regency wasn't very good. Poor technical quality with a $49.95 price tag, which in 1955 dollars was nearly a week's pay for many people, prompted Consumer Reports to warn in its April 1955 edition, "The consumer who has been waiting for transistor radios to appear would do well to wait for further developments before buying." The audio quality, especially of the human voice, left much to be desired. The hunt for a better transistor was on, and Philco's research made the necessary breakthrough in 1954 with the surface barrier transistor.

The Philco transistor was so good that it made high-speed computing a reality. A year later, the National Security Agency (NSA) sought out Philco to build a transistorized version of ERA's Atlas computer. Project SOLO, as it was called, pushed Philco into the computer business, and the resulting Transac S-1000. became the first transistorized computer. It was another computer first for Philadelphia. But, of course, as a top-secret NSA computer, it was not for the world to see. The first transistorized computer for public consumption popping onto the Philadelphia computer scene would be the Rosen-led team's Transac S-2000 in 1957.

Raytheon lumbered onto the scene producing the very tardy and ONR-sponsored RAYDAC (originally Raytheon's Project Hurricane Computer) of which only one was sold to the Navy. Raytheon then got together with Honeywell to form the Datamatic Corporation, from which Raytheon soon bowed out, leaving a Honeywell Datamatic division to produce a grand-sized dud called the DATAMATIC 1000.

As many as 26 private companies tried to get in on the early electronic computer industry. Most didn't make it. Even IBM had a tough time of it, but had monumental resources to wait for the right opportunity to appear. Few computer customers other than the federal government, large banks, insurers, and huge industrial companies could pony up the $1 million to $3 million in cash necessary to purchase one of the big, early computers.

IBM's angle from the beginning, as it had been with their card tabulating business, was to lease their machines. Many more companies felt better about $16,000 on a monthly lease than blowing the corporate treasury on an extravagant outright purchase. To manufacture computers and then to lease them required enormous upfront capital to hang in there waiting until the lease revenue began to flow.

IBM had the bucks. With its 701 already in the marketplace in 1953, IBM announced the coming of its competition killer the very same year, rolling out its first 650 business machine in 1954. It was pricier than all the rest at $3,200 a month (just about a year's wages in 1953), but it was popular. IBM figured to manufacture 50 of the 650s but ended up installing nearly 2,000. "IBM's position in the punch card field was a tremendous advantage for the 650; for hundreds of business organizations it seemed to be the next natural step."

A Cinderella company new to the computing scene was led by Ken Olsen, an original bright boy and Whirlwind engineer, who moved into an old mill in Maynard, Massachusetts, to form Digital Equipment Corporation (DEC). Olsen had built Whirlwind's Memory Test Computer (MTC) and had been assigned to help oversee IBM's production line for the SAGE AN/FSQ-7. It was no coincidence that when DEC shipped out the world's first modern minicomputer, it bore quite a striking family resemblance to Whirlwind and MTC technology.

On February 17, 1950, as George Valley introduced Jay Forrester to his Valley Committee, the decade-long storm and stress of digital computing was barely astir.

Forrester had before him a receptive, interested, even eager audience in the Valley Committee. It was an utterly pleasant change from openly contentious people with brows knitted in suspicion, as were those from the ONR and so many others. The committee was really pulling for Forrester and his machine to do the job at hand. There had not been such fresh air around since the summer interlude that had been the Moore School lectures.

The promise of those days in Philadelphia had come to pass; he and Bob Everett and the other bright boys had also come to pass. Life was good again. Maurice Wilkes said of them in his memoirs: "By the time I knew them they were pillars of digital orthodoxy." Pillars, yet. What a nice, new chunk of respectability had been hewn out of the former mountain of doubt. The boondoggle boys became the bright boys to nearly everyone.

Familiarity had even abbreviated electronic digital computer to that of just digital computer; and the death knell for analog machines was tolling ever more loudly. At the front door of 211 Massachusetts Avenue a sign now read, "The Digital Computer Laboratory," and Forrester was its director. They were ahead of the pack in the field of digital computing and moving further ahead all the time.

While most others spent the 1950s with a head stuck under the hood of their computing machines trying to crank them into existence, the bright boys were upright looking from machine to machine, and pondering the possibilities of networks.

Chapter Seven
Into the Great Wide Open

Janus, the twin-faced Roman god of doorways and new beginnings, looks both back at the past and forward into the future. Janus was how Maurice Wilkes had described Presper Eckert.

Wilkes was praising Eckert's intellect and industry for escorting the world through the doorway of digital computing with ENIAC. Much the same comparison can be made about Jay Forrester.

In 1953, Forrester snapped the first frame of his magnetic-core memory into Whirlwind. It was a new beginning. With it, he forever slammed shut the door on unreliable tube memory that had been the bane of all early computers. From bane to balm, shaky computer memory quickly became a thing of the past. Programmers gained confidence that their meticulous routines would run properly; companies gained confidence that calculations of their all-important business data were sound.

The enormity of such an event is near lost on modern computer users whose only concern with memory is having enough of it.

IBM's 701 is a telling example of a great machine with a very big memory problem. As Saul Rosen tells it, "The electrostatic storage system on the IBM 701 was very unreliable...the mean time between memory failure at 701 installations was often less than 20 minutes." The 701 outfitted with magnetic-core memory (renamed the 701 M) was a huge improvement. Redesigned and resold as the IBM 704, it "was quite outstanding for its time and achieved a near monopoly for IBM in the large-scale scientific computer field." In the 1950s, the advent of such trustworthy memory provoked a stampede for digital computing power.

(Detail) Light gun and SAGE tracking scope. (Courtesy of MITRE Corporation.)

Forrester's revolutionary invention was more than just a complete commercial success. It totally replaced skepticism with user confidence for an entire industry and would continue to do so for over 20 years thereafter.

When IBM received its first subcontract to convert Whirlwind technology into the AN/FSQ-7 in October 1952, it reaped a windfall of bright boy innovations to commercialize. IBM wasted little time in capitalizing on the technological advantage.

It not only produced the 704 but also began building the SABRE seat-reservation system for American Airlines, which became "the largest commercial real-time data processing system in the world." And there in the thick of it all from the very beginning was none other than Forrester's digital friend and advisor, Perry Crawford.

Crawford had retired from the Navy in 1952, and the very same year popped up at IBM as a member of its Product Planning group. He was the first to identify airline reservations as an ideal focus for SAGE technology.

Very much like SAGE with duplexed computers and a national grid of phone lines connecting nationwide control centers, SABRE connected 1,200 teletypes scattered all over the country to the airline's computing center north of New York City. Forty years later in his memoirs, Tom Watson Jr. acknowledged that the SAGE contract "enabled us to build highly automated factories ahead of everyone else, and to train thousands of new workers in electronics."

Bright boy technology, modern production plants, and a skilled work force complemented both IBM's huge financial clout and its reputation from the tabulating machine business. It was easy for IBM to gaze into a mirror and see Snow White smiling back.

Remington Rand was unhappy that the bright boys passed it over in favor of IBM. Remington Rand also, and with good reason, was a bit surprised as well, especially seeing that its UNIVAC was the "it" machine of data processing, and that its acquisition of ERA's Atlas technology gave it a leg up in scientific computing.

A memorandum from Forrester from November of 1952, explains the advantages of IBM over rival suitors Remington Rand, Raytheon, Bell Labs, and Sylvania. The selection committee, made up of Jay Forrester, Robert Everett, Norman Taylor, and C. Robert Wieser, had a clear view of the computing landscape of 1952, and saw things much the same as their contemporary, Saul Rosen.

IBM was the hands-down correct choice to entrust with the country's air defense computers. That IBM's Poughkeepsie plant was relatively close to Cambridge was also a good selling point. The proximity of IBM's branch office in Hartford, Connecticut, also had considerable travel-time merit. It was there, following quickly on the heels of the earlier subcontract, that beginning June 24, 1953, groups of IBM and Whirlwind engineers rendezvoused to hash out AN/FSQ-7 design problems.

At first testy and a bit combative, each side quickly settled in and came to understand and respect the other. The Hartford meetings, known to both groups as Project Grind, were important first steps in coming to decisions about every system and subsystem of the huge project before them.

George Valley and his Valley Committee were noticeably absent from the selection process, which says much about Valley's trust in Forrester and the bright boys' decision-making and managerial skills. Generally perceived as a bunch of cocky inventor types, the bright boys displayed considerable acumen in the very modern practice of outsourcing production.

They rode herd on IBM as the latter first produced two experimental AN/FSQ-7 computers, called XD-1 and XD-2—one remaining with IBM as a working prototype for its engineers and the other shipped to Massachusetts as the bright boys' prototype. From these two machines, and any ongoing engineering advances performed on them, would come all the other AN/FSQ-7s.

For two sets of engineers separated by hundreds of miles of landscape, a good set of blueprints was critical. Everett's pioneering work in the annotated logical diagrams of a computer's innards and connections, what today is called "block diagramming computer architecture," saved time, money, and anguish.

In the fall of 1953, Forrester and Everett then set up a Systems Office run by bright boy John Jacobs to oversee design control and responsibility for the experimental XD computers through to the finished AN/FSQ-7 computer system. IBM followed their lead, creating its own three-person Engineering Design Office to interface with Jacobs' group.

Necessary as well to the development process were accurate research documents to share between both teams of engineers. Peeling back the covers of Forrester's laboratory reveals an early penchant for painstakingly thorough documentation and communication of everything that went on, beginning in the early pre–Barta Building days and continuing onward into the AN/FSQ-7 and SAGE projects.

Since there were no existing publications on digital computing anywhere to be found, the bright boys published every scrap of their own
research as they went along, even if distribution was limited only to themselves. There was the seminal *Report R-127*, and the forward-looking *L-1*,
L-2, and *L-3* reports, but they hardly scratched the surface of the volumes
of paper produced. "There were progress reports, trip reports, conferences
and conference reports, master's theses stemming out of work done in the
laboratory, technical memoranda, and discussions with visiting experts
from industry, academe, and government."

Everything was categorized and filed as an M-Note, C-Note, E-Note,
or Bi-Weekly Report. This unorthodox, high level of communication
remained consistent from the skeleton of shelves that was early Whirlwind
to a continent filled with AN/FSQ-7s, all of it "word processed" by hand on
clunky typewriters and then duplicated on mimeo machines.

Forrester even conducted weekly affairs called Friday afternoon teas
where he'd speak informally one-on-one with individual researchers; these
grew to group teas as time went on and the bright boys expanded their ranks.
It was all about discipline in pursuit of reliability, which was the all-abiding
goal to which all early computing ventures aspired, yet few ever achieved.

Forrester and Everett demanded reliability. If an engineer worked himself into an electronic dead end, his own notebooks and the circulated
knowledge of others were his only salvation. There was no running to a
bookstore or library for the latest on digital circuit design and no cash to
redo something over and over again until it worked. And their machine
showed it: Whirlwind's reliability became legend in the industry.

Forrester's great advance was one of three momentous events from the
early 1950s that propelled computer technology into the future. Another
was the outbreak of the Korean War in June 1950.

Although Truman was still looking to cut defense spending by a half-
billion dollars as late as May 1950, the North Koreans would change all that
penny pinching in a hurry. The war sprung open the military's purse strings
from $13.5 billion to $48 billion (July 1951) and sent the Department of
Defense on a feeding frenzy for anything war-winning, especially electronics.

"The motivation was the fear that unless the United States engaged in a
militarized containment of Soviet power, the Korean War could be a prelude to a much wider conflict with the USSR."

Stalin's generals had drawn up the war plan, and the North Koreans
executed the plan perfectly, taking Seoul, the South Korean capital, in a

matter of days. The danger of a wider war was far from an exaggeration. According to historian and Stalin biographer Edvard Radzinski, Stalin told his close Politburo buddies, "We have the opportunity to create a communist Europe but we have to hurry." By hurry, he meant while the United States still had a demobilized military that had shrunk to pre–Pearl Harbor numbers and had its hands full of North Koreans. Before he could make good on his threat, Stalin died on March 5, 1953, from a brain hemorrhage, or as Radzinski contends, was poisoned.

Curiously, four months later, on July 27, 1953, the war that Stalin started and the war that the United States felt sure was a harbinger of a wider conflict ended with a cease-fire agreement between the United Nations and North Korea.

Truman believed that Stalin was heading toward a showdown. The previous January in an interview with Carleton Kent of the Washington Sun-Times, Truman said of Stalin's Korean adventure, "It's the greatest error he made in his whole career. If he hadn't made that mistake, we'd have done what we did after World War I: completely disarm. And it would have been a cinch for him to take over the European nations, one by one."

The Korean War, like all wars, especially modern ones, held great promise for technology. Technology would grow particularly prosperous in the 1950s serving an expanding military that was ramping up to spread itself all over the globe. As Charles Bohlen, a 1950s U.S. Foreign Service officer in Soviet affairs points out in his *Witness to History, 1929–1969*, because of post–World War II demobilization, the United States had military facilities only "in England, where we had transit privileges, and Saudi Arabia, where we had an airfield.

"As a result of our over interpretation of Communism's goals, we had by 1955 about 450 bases in 36 countries, and we were linked by political and military pacts with some 20 countries outside of Latin America. It was the Korean War and not World War II that made us a world military-political power." Or, from Truman's perspective, it was Joe Stalin who begat the U.S. colossus.

Although Korea was called the ugly war that nobody won, the real victor in the conflict was Japan. Historian Yoneyuki Sugita writes that the Korean War came along like a "divine wind" for Japan and its devastated economy. "Japan received large orders from the United States to manufacture military supplies and to repair ships, tanks, jeeps, aircraft, and others…returning Japan to a prewar productivity by 1951…and from a $300 million trade deficit in 1949 to a $40 million trade surplus by December of 1950."

Japan got acquainted with, and adapted well to, the war-imported assembly line and mass production techniques from the United States. Such an unprecedented opportunity to pull themselves out of an economic hell was not lost on the eager and industrious Japanese.

They created their own divine wind. An unknown company named Tokyo Tsushin Kogyo, Ltd., known today as the SONY Corporation, acquired Bell Laboratories' transistor patent in 1954; and by July 1956 Sigeru Takahashi and his mates at the Japanese government's Electro-technical Laboratory had built Japan's first transistor computer, the ETL Mark 3—barely a year after Bell Labs' own TRADIC (TRAnsistorized DIgital Computer). Japan's electronics industry was on the rise, and somewhere in there a little homage might be due Comrade Joe.

From 1950 to 1955 was a mere five years for 450-odd military bases to suddenly pop onto the face of the earth. Fast work, to say the least. And with those 450-odd bases came the need for expansive lines of military communications by radio, teletype, and telephone among far-flung bases and their headquarters, for airborne communication and navigation, for radar systems, and, yes, for computers…dependable computers that could do more than just count. SAGE-type networks would eventually find themselves watching European skies for NATO as well as Asian airspace from a newly remilitarized Japan.

Another ugly but important war took place in America's courtrooms. The Justice Department's antitrust actions against AT&T and RCA in 1949, and then with IBM in 1952, each resulted in consent decrees in 1956 that blew up their high-tech patent monopolies. No one demobilized after this war. The big three bowed low to the court but then regrouped in subtler ways for a quick return to business as usual.

Time was of the essence for the smaller and newer electronics wannabes. AT&T was allowed to stay only in the field of telecommunications and had to license its patents. As for RCA, "[t]rustbusters," reported Time, "charge that the company uses its pool of 10,000 radio-electronic patents to keep industry in the dark about new developments, and forces licensees to pay for RCA patents they do not need."

IBM was dealt a huge blow. Its former practice of only leasing a machine, and never selling one, was struck down by the Southern District Court of New York. An IBM 700 Series computer that formerly leased for $30,000 a month could be bought for $1,700,000.

Worse yet for Big Blue, it could not hold customers in bondage to service and parts available only through IBM. Services and parts could in many cases add an extra $4,000 a month to a computer lease. If the mean time to failure on an IBM 701 was 20 minutes, then parts and services could well be a fantastically lucrative business. The 1956 decree required IBM to train personnel and to provide technical manuals for anyone who owns, repairs, maintains, or distributes IBM electronic data processing equipment.

And the decree also ordered IBM to grant unrestricted licenses to companies desiring to use its patents, and enjoined IBM from instituting any suits for patent infringement that occurred prior to the date of the decree. The end result of the consent decrees was that they afforded the little guy a slim chance to get into the game before the door slammed shut again, which it did but not before they did.

Forrester's magnetic-core memory was swept into the IBM black hole of patent control. IBM would have been insane not to try to control in some manner, shape, or form Forrester's invention. Using magnetic-core memory in SAGE government computers was legal, but its use for commercial purposes was not, which IBM desperately needed to do for replacement of unreliable electrostatic memory. The company turned to its Intellectual Property Department, which kept a worldwide patent watch for any infringer or potential patent menace to its stable of inventions.

Run by an IBM loyalist of 25 years named James Birkenstock, the Intellectual Property Department took considerable interest in the invention of its newest best friend from the Barta Building. Birkenstock and his department plus IBM intellectual property lawyers swept the world clean of any opposition, potential opposition, or even any seemingly insignificant hint of opposition to IBM's commercial use of magnetic-core memory.

Armed with cash and clout and a smile, Birkenstock swept away potential threats—a pile of cash went to a Los Angeles public works inspector who invented out of his basement laboratory; An Wang got $500,000; An Wang's codeveloper Way-Dong Woo was bought off; and a German named Gerhard Dirks got $1,000,000 plus IBM's help emigrating to the United States with his wife, and a job in sunny California at IBM's San Jose plant.

IBM's own Munroe (Mike) Haynes attempted to put his core memory design into the mix, but it was scrapped by IBM as unworkable. And anyway, why waste the effort when Forrester's prize was there for the taking? The only possible threat to IBM's use of core memory came from RCA engineer Jan Rajchman, who had a magnetic-core memory patent pending at the U.S. Patent Office.

Since Birkenstock and IBM had a previous cross-license agreement in place with RCA, IBM was free to use the technology. Tom Watson Jr. ordered his company to discard all electrostatic memory for the 700 series and retrofit them with core memory.

But what about Forrester and MIT? They were in line to get zero. IBM begged innocence with the limp excuse that Rajchman was first in line at the Patent Office (1950), Forrester second (1951), and that was that. The waiting was over in 1955 when Research Corporation, the Cambridge patent licensing agent for MIT, filed suit against IBM for "willful infringement of the Forrester patent"—supposedly without informing MIT it was doing so.

Tom Watson Jr., who sat on MIT's board, promptly resigned his MIT seat; James Killian, MIT's president, who sat on IBM's board promptly resigned his IBM seat in return. And the royalty that Research Corporation sought from Birkenstock and his crew of patent sweepers was considered by IBM as outrageously greedy.

Something had to be done; no one at IBM dared rock the boat. Watson and Killian settled it old-boy style: IBM paid MIT a lump sum of $13 million; MIT paid off Research Corporation, and then promptly fired them. IBM would go on to produce trillions of ferrite cores and use them to make millions of magnetic-core memory planes. Birkenstock secured a fortune for his company, and IBM would rake in profits hand over fist until 1970.

Later, during litigation, Rajchman admitted to Forrester's prior discovery of magnetic-core memory and also conceded that Forrester had been first to develop working core memory and first to install and run it in a computer.

However, Forrester had been late in filing his claim within the prescribed limits set by patent law. With all the paper published by the bright boys on everything and anything at their laboratory, somehow—incredibly—this "slight" paper-filing oversight had taken place. Somehow too, Research Corporation had been less than diligent in failing to notice the elephant in the room. How could they have missed it?

In 1956, Jay Forrester finally prevailed in court and toted home his long-overdue patent: no. 2,736,880. First the SAGE contract, then Whirlwind technology, and finally core memory; the bright boys were IBM's "divine wind."

Between 1954 and 1956, IBM's gross income soared from $570 million to $892 million; it added 22,000 new employees, split its stock five-for-four, and declared a 2.5 percent stock dividend. Consent decree or no, IBM was virtually unstoppable.

By 1956, IBM was no longer a tabulator company but "the world's largest and most profitable computer manufacturer, building the machines by the hundreds." The same "divine wind" blew MIT's way as well: "Royalties to MIT from non-government sales amounted to $25 million, as MIT licensed the technology broadly."

In 1970, up popped a newly formed company calling itself the Intel Corporation. Intel released its 1103 transistor chip, the first DRAM (Dynamic Random Access Memory) chip, which deep down inside was quite akin to Forrester's core memory concept but made really, really, really small. By 1972, the 1103 was the bestselling semiconductor memory chip in the world.

In the early spring of 1950, as Forrester worked with the Valley Committee and watched their faces daily growing more confident and enthusiastic about his mammoth computer, he grew less confident.

George Valley, Director of SAGE, 1952–1958, speaking to members of the press at the announcement of the SAGE system for continental air defense. (Courtesy of MITRE Corporation.)

He knew full well that the electrostatic tube shop in the Barta Building was an exercise in futility. But he had to keep up a good front and wait. As late as 18 months after his first introduction to the Valley Committee, August of 1951, his lab's snappy-looking Whirlwind I brochure dedicated a full page of praise for electrostatic storage. Complete with a photo layout of tube makers in white shirts, ties, and surgical gloves handling tube assemblies with medical instruments, the brochure's glossy outward appearance belied the undercurrent of unease that preoccupied Forrester.

He also knew full well that when he fitted Whirlwind with its first set of memory tubes, they were doomed to early burnout. Reliability meant only one thing: magnetic-core memory. Core memory existed only in the pages of his notebooks.

Everything that the bright boys had worked six years to achieve was riding on those notebooks, as was George Valley's reputation, as were the hopes of Muir Fairchild and Gordon Saville's air defense, as was the rapid growth of a young computer industry. How could so much be riding on one man's notebooks? A new decade was dawning. Joe Stalin was ready to play his role, as was the Justice Department.

What about Forrester?

Lincoln Laboratory

George Valley's encounters were unusually interesting and seemed always to have surprising impact and consequence well beyond their intended purpose. One of his most conspicuous encounters was a December 15, 1950, luncheon with Louis Ridenour at the Chief of Staff's table at the Pentagon. Valley was in Washington on other business when Ridenour passed a note to him requesting a rendezvous.

Valley and Ridenour were not strangers. Much to the contrary, they were friends; both were Rad Lab grads and members of von Karman's Air Force Scientific Advisory Board (SAB).

Handsome, a meticulous dresser, athletically fit with a rugged edge, Ridenour looked none the part of the brilliant physicist that he was. A professor at the University of Illinois and former overall editor of the Radiation Laboratory series on which Valley had also labored, Ridenour was in 1950 the first chief scientist of the Air Force. It was as both friend and chief scientist that Ridenour asked Valley to lunch. He had a favor to ask, a proposal to make, and some advice to share.

Valley wasn't completely in the dark about Ridenour's purpose for the meeting. He was well aware of the traffic of interest afoot regarding an expansion of the Valley Committee. He had already chatted with some ex–Rad Lab veterans, who had signed on with Air Force Headquarters and were hoping to resurrect a semblance of the Rad Lab, especially to counter a rising Soviet threat.

The raging Asian war that by December was six months on and not going well was of particular importance. The Inchon landing had taken place the previous September, and the North Koreans looked defeated. But the Chinese suddenly entered the war and Russian pilots were flying MIGs.

As the two men met for lunch, the First Marines were retreating from the frozen Chosin Reservoir to a port of escape at Hungnam, 78 deadly miles to the south. Missing from the war was an OSRD-type organization, the kind Vannevar Bush had advocated in *Science: The Endless Frontier*, writing "our defense against aggression demands new knowledge. This essential new knowledge can be obtained only through basic scientific research."

The Air Force and their chief scientist were trying to jumpstart their new-knowledge machinery of basic research. A month after the war began, the Air Force, tipping its hat to Valley, mirrored the Valley Committee by starting a similar committee to review the guided-missile program. At the time, Ridenour chaired an SAB committee putting together the new Air Research Development Command (ARDC) where there was strong interest in setting up a laboratory for basic research and development.

By May of 1951 ARDC would become a major Air Force command under Major General David Schlatter, and would be looking for a steady supply of modern electronics for command and control of air defense, SAC, tactical operations, and the brace of new air bases around the world. All of this was very much in keeping with von Karman's R&D master plan in *Science: The Key to Air Supremacy, and Where We Stand*, all of which brought George Valley quickly front and center into the eye of the movement.

Valley and Ridenour met on a Friday, a week before Christmas, a good time to reflect on the remarkable accomplishments that Valley's team had rung up just short of one year since the Valley Committee began on January 20, 1950.

One of the moving parties for the new laboratory was Ivan Getting (then Air Staff Assistant for Development Planning), who admired the way that the Valley Committee cut to the chase when applying evolving technology to meet Air Force operational needs. It was very Rad Lab–esque to Getting and others.

The Valley Committee "was not inhibited by the bureaucratic inhibitions and restraints characteristic of military organizations." As Valley put it, his old Rad Lab mates "demanded" a show-and-tell of the Valley Committee's air defense toys. They liked what they saw and came away convinced that a new Air Force–financed air defense laboratory was necessary.

Getting made his position clear to both Valley and Ridenour. In October, MIT's Stratton related to Killian a conversation with Ridenour in which Valley's work was spoken of as "our brightest hope in the field of air defense," and that there was a need of a laboratory to further that work. A month before the Pentagon lunch, November 20, 1950, Ridenour sent a memorandum to General Saville requesting that he move on the air defense laboratory "by negotiating a research contract with a suitable institution in the Cambridge area…and MIT has indicated that they would consider taking such a contract."

The memorandum, titled "Proposed Augmentation of ADSEC Activities," called for a laboratory of 100 researchers and a budget of about $2 million a year. Valley had been copied on the memorandum. Lunch would be long, and Ridenour wanted to get a personal feel for Valley and his committee's work. Successes were everywhere for Valley's committee, so it was easy to give Ridenour the cook's tour. Ridenour already knew most of what Valley would tell him, but that is where the favor and the proposal came in. Ridenour was there to draw out Valley's commitment to an ongoing leadership role in air defense after the Valley Committee came to an end.

The success of the September radar tests was critically important. Charles Draper's Instrumentation Laboratory had a hangar at Hanscom Field with a

Lincoln Laboratory (Building F blockhouse, center left), 1955. (Courtesy of MITRE Corporation.)

pilot and a B-26 for use in the tests. "Harrington and Forrester demonstrated that radar data could indeed be received from a phone line, manipulated by a digital computer, and then displayed on a cathode ray tube." To Valley, the test was "proof of principle, and sufficient for ADSEC's purpose, which was to tell the Air Force what to do, not actually to do it for them."

What he had signed on for with Robertson during the taxi ride was now delivered, and as such, Valley could argue that his mission was at an end. Returning to cosmic ray studies and the good life was his for the asking. But he did not ask. And Ridenour was not about to offer it. Ridenour wanted more than proof of principle; he wanted an air defense system, and he wanted Valley and his boys to make it happen.

Valley was willing to agree with Ridenour in large part because of the early flak such a notion had already stirred up at MIT. "Some decided to oppose the setting up of a new laboratory unless they could run it," then a "second group ... began to attack ADSEC and me as incompetent."

"Jerome Wiesner and Jerold Zacharias," wrote Ivan Getting, "opposed the venture out of fear that it might dilute their own RLE [Research Laboratory of Electronics] programs." Gordon Brown feared that RLE's work would atrophy, and the Army and Navy "were fearful that their influence with MIT would be eroded."

When Ridenour asked Valley to draft a letter to MIT's president, James Killian, asking for just such an air defense laboratory, Valley was quick to oblige. "I completed it in about an hour, and Ridenour spent another 15 minutes recasting it into appropriate general officer's diction." Ridenour had it typed, and then got Vandenberg's signature on it by four o'clock that afternoon. The Valley-Ridenour-Vandenberg letter went that same day directly to Killian: "The Air Force feels that it is now time to implement the work of the part-time ADSEC group by setting up a laboratory which will devote itself intensively to air defense problems."

What had been merely a "notion" before lunch was now damn close to a command from the highest officer in the Air Force and one of the most powerful men in Washington, whose F-86 pilots were tangling daily with MIG-15s over Korea.

Air defense was big stuff, bigger than MIT and Killian. Senator Henry Cabot Lodge Jr. told his fellows on Capitol Hill what they already knew, that the air defense of America was "so feeble as to almost invite attack."

Diatribe or not, MIT was not going to buck that current. Financially, morally, and academically MIT was left without a scintilla of choice. Before putting the university's official imprimatur on the Vandenberg request, MIT first needed a little face saving, to placate its upstaged military sponsors, and to cheer up unhappy researchers. MIT asked former Rad Lab second-in-command, F. Wheeler Loomis, to come in from the University of Illinois to head up a study group that would assess ADSEC's work and make recommendations. It all took shape in February 1951.

Occupying the upper floors of the recently purchased Lever Brothers Building, in space scheduled for MIT's new School of Management, the well-respected Loomis kicked off the examination of the ADSEC extension into a permanent air defense laboratory.

Project Charles—as it was called because the Lever Building overlooked the Charles River—consisted of 28 full-time scientists and engineers; 16 consultants; air officers from Great Britain, Canada, and the United States; 2 business managers; 18 secretaries; a librarian; a mechanical engineer; and 2 assistants. Among the group leaders selected for Project Charles were Wiesner, Zacharias, and Gordon Brown. Attending, as two of the consultants, were Ridenour and von Neumann. For six months, Project Charles closely examined all findings. Valley wrote everything into its final report—in which there was not one proposal to improve upon the original ADSEC recommendations.

On April 20, 1951, the bright boys tossed the clincher at Project Charles. With Forrester's first set of electrostatic tubes in Whirlwind, they performed the first-ever air detection, tracking, and interception—not once but twice more for good measure. Three times in one day. With the B-26 as the interceptor, the bright boys and Whirlwind tracked and scanned a T-6 trainer and a Beechcraft twin-engine C-45, each time maneuvering the B-26 to within 500 yards of the targets.

Barta Building, Room 222, circa 1954. (Courtesy of MITRE Corporation.)

Whirlwind's detractors were totally overwhelmed by the air defense display. "Much of the credit goes personally to Jay Forrester," said Valley. "Each day, after enduring hours of Project Charles, he nursed his balky storage tubes late into the night."

Project Charles reaffirmed ADSEC. All that remained was to divvy up the new prize and give it a name. "At MIT's insistence ... the three services contributed to the budget in roughly the following proportions: Army, Navy, Air Force: 1, 1, 10.

In this way, the Air Force was allotted most of the services of the new laboratory (and paid the most) ... the opposition group of scientists were granted a dominant influence in the laboratory, because most of the directorships went to them, as well as control of the Army and Navy money."

Project Charles concluded warmly for everyone on Winter Place at Boston's Locke-Ober Restaurant over dinner and a celebratory toast of Napoleon brandy.

At the nexus of Lexington, Bedford, and Lincoln, Massachusetts, a patch of land near Hanscom Field was selected as the site for the new air defense laboratory. Since MIT already had a Project Lexington and a Project Bedford, the freshest name left was Project Lincoln. As bulldozers ground into action and clawed away at Lexington's forest clearing sites for buildings, the name Lincoln Laboratory was selected to go above the front door. The bright boys had outgrown Cambridge and would now have a complex of buildings and their own idyllic chunk of suburbia.

Sons of Whirlwind

Project Lincoln got rolling in Cambridge in July 1951; the official Air Force contract was signed in September. Buildings were promised to be ready in the Lexington woods by April 1952. Awaiting the move and collocating together on campus were the elemental parts of the air defense laboratory: radio communications, long-range communications, radar, solid-state physics, airborne early warning, systems components, and ordnance. One thing was for sure in the transition: whenever and wherever the bright boys moved, Whirlwind was not going along with them. Whirlwind was wonderful but, as with many inventions, it was a one-of-a-kind machine that was impossible to mass-produce. The boys needed a production prototype of Whirlwind that a factory could better duplicate and manufacture.

The bright boys immediately began to plan for such a machine, one they called Whirlwind II. Whirlwind II, however, would never happen. Project High came along in the fall of 1952, making IBM the provider of a reproducible Whirlwind. The new Whirlwind would come as the identical XD twins: XD-1 going east for the bright boys and XD-2 staying put in Poughkeepsie. When production of these prototypes bogged down, Project Grind ground into being in June of 1953 to "unbog" Poughkeepsie.

Forrester and Everett used the meetings to get directly to "the technical design problems, making it clear circuit by circuit and system by system" what IBM's engineers needed to do. A delivery date for XD-1 was set for January 1, 1955; a building to house XD-1 was also planned for Lexington, which would be named Building F, with an anticipated completion date also slated for January of 1955.

Still another name change took place when XD-1 took on its official military nomenclature: AN/FSQ-7. The new name proved to be a mouthful, so it was nicknamed the FSQ-7, then shortened again to just the Q-7. Since the new machine would not be around until 1955, old faithful in the Barta Building would pull the load until then. Project Lincoln began moving into its new street address at Lincoln Laboratory beginning in the spring of 1952, continuing into 1953, and completing in 1954. Left behind in the wake, the bright boys began readying for their biggest test yet, the famous Cape Cod System.

Another smaller Whirlwind, called Whirlwind 1½ or sometimes called Whirlwind 1A would go into development on nearby Vassar Street in MIT's Whitemore Building, formerly the Whitemore Shoe Polish Company. Ultimately, this machine also underwent a name change, becoming known as the Memory Test Computer or MTC. Ken Olsen, a former graduate student of Forrester's, who like Bill Papian wrote his master's thesis on magnetic memory, headed the MTC design team.

Assisting on the MTC design was a Forrester new hire (1951) and newly anointed bright boy, Wesley Clark, fresh in from the West Coast where he'd been working at Hanford on atomic reactor research. Clark soon proved to be another of Forrester's more prescient employee placements: he and Olsen would later migrate MTC technology to Lincoln Laboratory, where they would marry it all to transistors and a bit of new machine design to produce two remarkable computers, the TX-0 and TX-2.

"It was supposed to be an honest-to-goodness computer that would really run and test the memory, but not a computer that was designed to be useful," remembered Olsen about his days constructing the MTC. "I was given the job of building the computer just as soon as my thesis was done. I think I was still a graduate student and it cost a million dollars. I can remember being impressed how much a million dollars was. How much work it took to spend a million dollars."

If and when Forrester got his magnetic-core memory together, the MTC would be used to test how well it worked or didn't work. If all went according to plan, electrostatic storage would be removed from Whirlwind and replaced with core memory. As Olsen remembers, within one day of the MTC cranking into action in May of 1953, the electrostatic storage tube facility was shut down and everything switched to core memory.

Bright boy Pat Youtz, who had led the tube laboratory, immediately switched over to leading the effort for Whirlwind's many display screens in the Barta Building's top-secret Room 222.

Even with the MTC's success, Olsen was not done with the machine: he and Clark got the chance to build it again—this time with transistors—in an isolated basement in Lincoln Laboratory.

Fueling their ability to evolve the Memory Test Computer through to the newly emerging world of transistors was the insightful creation of Group 63, the Advanced Development Group at Lincoln Laboratory. Instead of cloistering all research in Building F as military, and therefore classified, the bright boys reached out to innovation once again.

"When the Digital Computer Laboratory was formally made Lincoln Lab's Division 6 for classified work under the SAGE umbrella," remembers Wes Clark, "Forrester and Everett saw to it that basic, unclassified research and development of computers would not simply disappear in the process." To that end, they organized Group 63. Initially under David Brown's leadership, and later Bill Papian, Group 63 brought together those who had worked on the MTC plus others doing research and development on memory and circuits. There were even subgroups with physicists, chemists, and other engineers as team leads.

There was also a practical side to the creation of Group 63: survival. The concern was "that SAGE was consuming Division 6," recalls Bob Everett, "and if we did not do something, we would end up with no future work at all. We, therefore, set aside a selected group of staff with the mission of continuing basic work on computer technology and protected it from the demands of SAGE. SAGE provided the money."

With SAGE as paymaster, as many as 30 staff then had the chance to forge ahead with the likes of "transistor machines, core memories, time-sharing, computer games, printers, and many other things," safeguarding a future in basic research while keeping apace or even keeping ahead of ever-newer breakthroughs in technology—all of which was beneficial to both SAGE and Group 63. Much of their work would inure to the paymaster as payback, as with "the 65,000-word memory that was of great value to SAGE."

As things would eventually turn out, this insightful creation would prove to be a wonder box of unintended consequences. With Clark handling the logic design and Olsen the transistor circuits and hardware, they proposed a new machine that they tabbed the TX-1.

"The idea was soundly rejected by Jay and Bob Everett," says Clark, "on the grounds that we didn't yet know enough for such a big step." Back to the drawing boards they went, this time proposing what Clark refers to as a "primitively simple computer." This new-age MTC would go by the name of TX-0 (the TX referring to Transistor eXperimental, and pronounced as "Tixo"). Olsen, naturally influenced by Whirlwind and the MTC's packaging, circuits, and toggle switches, rebuilt MTC's circuitry using 3,600 of Philco's surface barrier transistors at $80 a pop ($288,000).

The TX-0 turned out to be a 5-MHz, parallel, general-purpose, stored-program, digital machine. It had a cycle time of six microseconds and was capable of performing better than 80,000 additions per second. One objective in building the machine was to test and evaluate the use of transistors as the logical elements of a high-speed computer. The second purpose was to provide means for testing a large capacity, 65,000-word magnetic-core memory.

The TX-0 (1956) was a success that spawned in 1958 an even larger, more advanced descendant in the TX-2 (22,000 transistors). Wes Clark and his team designed the TX-2 for Ken Olsen's team to build. It was this TX-2—with Larry Roberts at the helm on the East Coast, networking westward to Tom Marill in Santa Monica at the helm of another Whirlwind successor, the AN/FSQ-32 (1965)—that was the original transcontinental hookup that heralded the beginning of ARPANET. "This demonstrated that a packet-switching network could work across the country, linking two different computers running two different operating systems, thereby making a key step in the evolution of what became the Internet."

The TX-0 would serve Ken Olsen again. Impatient for the TX-2's completion, which wouldn't happen until 1958, Olsen left Lincoln Laboratory in 1957 and founded the Digital Equipment Corporation (DEC). His early line of PDP computers were reported to be advanced versions of the TX-0 wearing DEC logos. Digital, as Olsen's company was known, eventually became the near equal of IBM.

These sons of Whirlwind—MTC, XD-1, AN/FSQ-7, AN/FSQ-32, TX-0, and TX-2—moved seemingly in every direction at once: changing their names, rearranging their form, physically moved, and re-purposed, yet still influencing everything in their paths.

Meeting notes from Group 63's Steering Committee from June of 1958 display the wide diversity of intended research: "data processing, pattern recognition, voice recognition and other self-organizing system studies," which even included a stint with Charles Molner of the Communications Biophysics Laboratory under Professor Walter Rosenblith, "to aid in the analysis of brain-wave data gathered from the auditory cortex of a cat's brain."

Democratizing the Computer

However, the true sons—and daughters—of Whirlwind weren't metal, glass, magnetic cores, or even transistors, but very human.

Overjoyed that Whirlwind was staying put in Cambridge were the hundreds who journeyed to the oracle for help, got it, and went away true believers. This diaspora of the "converted" created a large, ever growing, and enormously influential fan club that proselytized the great machine far and wide.

In the 1950s, if anyone wanted to know which way to the future, all one had to do was to follow graduate students and others heading over to the ever-popular 211 Massachusetts Avenue.

Across the MIT campus and beyond, researchers from academia, business, and industry lined up to get their calculations done. If a problem was either too time consuming to do by hand or totally intractable except by machine, Whirlwind would step in to crunch the numbers. With plenty of calculations in need of such treatment, Whirlwind was not only in demand but, as would be expected, became an admired member of everyone's family.

Although working on top-secret government air defense projects, the bright boys offered machine time on Whirlwind to anyone. Even "senior mathematicians had lost interest in the application of analog computers for analyzing random processes. They were switching to digital techniques," and although "Whirlwind was overcrowded, almost no one wanted to use the much slower and less accurate Differential Analyzer, so it was scrapped."

The bright boys were putting the right tool into the right hands, and in doing so, created a digital movement.

An MIT grad student, Eldon Hall, and his wife were two of many. Hall, who years later would design the Polaris missile guidance system and the Apollo Guidance Computer that landed astronauts on the Moon, got his first look at a digital computer in 1952, watching Whirlwind compute the antenna pattern for a dipole antenna and display the results on an oscilloscope. "I learned that this monstrous marvel was available free of charge to researchers with complicated problems."

His wife, needing to compute the eigenvalues of matrices, asked the great oracle for help. "My wife submitted the elements of a matrix to an individual at the Barta Building. He reappeared a few days later with the results, truly a marvel when compared with the hours of labor using a mechanical calculator." Hall was hooked.

Later, when the Air Force required the extreme accuracy of a digital computer for intercontinental missiles, and Hall's lab "had no experience in digital control systems," Hall began building them.

For 22-year-old Enders Robinson, the allure of Whirlwind was extraordinarily fascinating and would lead him into a new career, one where he would revolutionize the field of geophysical exploration—basically, the search for oil.

In 1952, as an assistant to MIT mathematician G. P. Wadsworth, Robinson was given eight seismic records from Magnolia Oil, the research arm of Mobil Oil, and asked to submit them to some, any kind, of mathematical scrutiny. Just what sort of scrutiny no one had the foggiest idea.

"The first time I ever saw a seismogram," he said years later in an interview, "it was not a pleasant feeling to look at that jumble of lines crisscrossing all over the place and know that you're going to have to try to find a pattern in the mess. In those early days, I had no idea about how oil was found and it wasn't an easy thing to pick up in Boston. I went to the library but at that time there were only three books on geophysical exploration. After reading them I felt worse off because I realized how little I knew."

Like so many others searching for answers to tough questions, Robinson traipsed on over to the oracle in Forrester's Digital Computer Laboratory. Robinson's basic plan was to take the seismic records and apply to them mathematical methods based on Norbert Wiener's concepts for deconvolution and time-series analysis, which came from Wiener's recently published book from 1949, *Extrapolation, Interpolation, and Smoothing of Stationary Time Series.*

Deconvolution is the algorithm-based process used in the techniques of signal processing and image processing. "In the spring of 1952, I went to Whirlwind with Howard Briscoe and put deconvolution on Whirlwind." And with that first brief encounter, the great machine quickly added yet another new convert to its ranks. "Whirlwind was good for signal processing because it could handle a lot of data … for signal processing, for deconvolution … it was very fast and efficient … it could handle a lot of data that the other machines of the early 1950s could not."

For the young mathematician, the encounter was nothing short of life changing. He had hit upon seismic research's core concept: converting continuous seismic traces into digital form and then using mathematical methods to enhance the data. Soon after, embarked on a new career as a geoscientist, he next founded the Geophysical Analysis Group at MIT, which a decade later led to a digital revolution in geophysics. Today, his book, *Digital Imaging and Deconvolution: The ABCs of Seismic Exploration and Processing*, co-authored with long-time colleague, Sven Treitel, is still a must-have resource for every geoscientist.

These advances in digital imaging and deconvolution eventually found ways into yet other areas of observational science, such as oceanography, meteorology, planetary science, astronomy, and medicine—in fact, medical ultrasonic imaging is the direct counterpart of seismic imaging.

Although Projects Charles and Lincoln were soon presented the looming reality of trucking everything off to the suburban retreat that would become Lincoln Laboratory, the great machine was staying put, and Forrester and Everett made certain that research applications using Whirlwind's computing power were never abandoned at 211 Massachusetts Avenue.

To the contrary, the pace in the Barta Building was near furious. In the four months between December 1952 and the end of March 1953, a total of 781 programs (exclusive of military applications) were run on Whirlwind. Nearly 200 a month, which was even more remarkable given that military use of Whirlwind was a priority.

"The military took about eight hours a day," recalled Robinson, "and since it was a vacuum tube machine, the maintenance took about eight hours a day, and the academic people were supposed to get eight hours a day, but it never worked out that way because sometimes the maintenance took more time."

Put in charge of squeezing non-military applications into these eight hours or less was the energetic and ever resourceful bright boy, Charles Adams. In 1953, the 28-year-old Adams became one of the first systems programmers in the Digital Computer Laboratory's Mathematics group. Adams was particularly concerned with preparing routines for Whirlwind, including the assembly language program, the design of floating-point arithmetic, and utility routines for debugging. But it was as the head of the Scientific Engineering Applications group that he provided invaluable assistance to those eagerly awaiting a turn at solving unclassified problems in physics, chemistry, or engineering, or those from most any other discipline with the yen to take a dive into Whirlwind's world.

The gamut of problems that Whirlwind was asked to unravel would include not only those as in Robinson's deconvolutions but also excursions into areas like deuteron binding energy and wave functions for MIT's Physics Department; or transient aerodynamic heating of flat plates for the aero-elastic and structures laboratory; or optical properties of thin metal films for the Chemistry Department; or even writing a graphics program together with the visiting Maurice Wilkes, creator of EDSAC.

Whirlwind was a magnet for those searching for answers to what mathematicians call "hard" problems: those intractable problems that throw up roadblocks to progress somewhere, sometime, in most every discipline.

And Adams was the master guide, busily leading one group or individual after another through Whirlwind, helping each to open doors of perception into worlds previously thought inaccessible. And, most often, succeeding.

For those a bit uncomfortable, unfamiliar, or downright skittish around the new field of electronic, digital computing, Adams was also there to teach the tricks of the trade to any (including military personnel) who wished to know. He taught a monthly programming course in MIT's Electrical Engineering Department using the Maurice Wilkes, D. J. Wheeler, and Stanley Gill book on programming the EDSAC, as well as a course on advanced programming techniques.

In the summer he ran an intensive, two-week program on computers and computer applications that included an introduction to digital-computer coding, a survey of existing computers, applications, numerical methods, and advanced programming techniques, all supplemented by group discussions and by demonstrations and practice on Whirlwind.

Ever the stylish promoters of the new digital movement, Forrester's Digital Computer Laboratory not only put out flashy Whirlwind brochures but also made a Whirlwind movie in 1953 titled *Making Electrons Count*. Part of the movie actually depicts how the typical supplicant approached the oracle. A physicist from the Retina Foundation comes with optical design formulas that Whirlwind successfully crunches for him. The film's voice-over narration then brags: "We have told you the story of one problem recently solved on one digital computer … you have an idea of the current importance of the digital computer as a new tool to help scale some of the hitherto insurmountable peaks which span the domain of man's activities."

"Hitherto insurmountable peaks" may have been a little over the top, even for 1953, but that's exactly what pilgrims sought and got from Whirlwind.

For one mathematician who made a pilgrimage to the oracle, just dropping off a problem and then returning the next day for an answer was not enough. Douglas Ross wanted a more hands-on experience with man and machine working out the problem together. At the time, such thinking was a bit fanciful. Most people were thrilled to death at handing over a stack of IBM cards for an answer the next day.

"The next day you got your answer back," remembered Olsen, "and it usually was that you'd made a mistake," which meant starting all over again with IBM cards.

Ross, who came to MIT in 1951 as a freshman calculus instructor, saw Whirlwind's consoles and display scopes for the first time in the summer of 1952. "Significant events," he later recalled, "always are the rearrangement into new forms of otherwise ordinary and insignificant, routine happenings."

The Whirlwind viewing screens that were used for radar tracking and plotting coordinates of aircraft, Ross saw as personal workstations for interacting with a computer to solve problems or to accomplish a task. If something was wrong, then you and the computer worked it out right then and there.

Which, of course, was exactly what J. C. R. Licklider contemplated in 1960 in his famous "Man–Computer Symbiosis," and what Doug Engelbart also in the 1960s saw as "human augmentation." Ross was ahead of both by a nearly a decade. Or was it the machine, Whirlwind, that prompted such radical thinking?

As Engelbart articulated it: "It's not just about putting a person in front of a computer. Computers are just a central part of the tool system. It's the combination of a tool system and human systems working in concert which is really powerful."

Bright boy Jack Arnow introduced the young mathematician to the wonders of programming Whirlwind, and Ross quickly took it from there. By 1953, he had gained access to Whirlwind's secret room: "the locked, green double doors of Room 222," with all the air defense consoles, especially the E31 Console, the master control station that, if necessary, could intervene and run the entire machine. The all-black room with red light from indirect ceiling lamps was eerily fascinating for him. So too was the large wall image of an outline map showing the New England coastline; it was glowing at him from a screen and light canon of what was called the Area Discriminator.

He wrote new code, devising programs to harness the huge machine to interact with him at a console. His purpose in Room 222 was to use Whirlwind to evaluate the performance of airborne fire-control systems, specifically, servo-controlled tail turrets in bombers. His fascination led him to use his experiences in Room 222 to build the first interactive, personal workstation in 1957 at Eglin AFB, Florida. During one typical adventure in Room 222 in 1954, Ross recalls the intercom crackling on and a voice booming out asking what he was doing alone in the darkened room. Ross had written a 200-line program to find and track a moving shadow. Hunched in a corner with his index finger pressed against a glass oscilloscope screen, Ross, his fingertip a blue-white spot of light, was writing his name "Doug," which was also writing itself on a screen in the Test Control Room down the hallway. It was the first-ever hand-drawn input into a computer.

Cheerios

From February 1950 to May 1953, Forrester had his hands full of memory. He had bottles of it that he did not want but had to use, and cores of it that he could not use but wanted to. He had been filling up page upon page in his notebooks on magnetic memory since 1949; he had seen An Wang's method; he had submitted a paper on magnetic cores to the *Journal of Applied Physics* in June 1950; and he had applied for a patent on "Multi-Coordinate Digital Storage" on May 11, 1951. The same day that he was filling out patent papers, the best he could do for his beloved machine was 16 bottles of electrostatic storage.

Forrester, with a bunch of Air Force money in his pocket, had Ernst Albers-Schönberg cooking up chemicals with the hope of finding the perfect core formula, and he backed that up with experimental core research at Arthur von Hippel's laboratory at MIT.

He had neither bottles nor cores ready for Harrington's first capture and transmission of pings from the B-26. He had to dance with only one set of bottles in Whirlwind for the famous Project Charles demonstration. In 1952, he ordered the building of the Memory Test Computer to test memory cores that might not be available to test.

He watched as Project Lincoln went from forest to concrete buildings, an entire facility waiting on a computer needing core memory for a national air defense system. He heard the gossip from "analogers," as Valley called them, who jeered from the sidelines hooting for failure.

As 1953 rolled into view, he had another major worry to add to a growing tale of woe: time was growing near for the anticipated Cape Cod System, when Whirlwind would be called upon to perform in an experimental air defense umbrella over a large chunk of New England.

If he failed to deliver the goods, all the Air Force money in the world wouldn't be enough to cover his tracks out of town. And if his own spirit was not dispirited enough from all the growing concern, he still had to keep everyone else motivated with uplifting talk of an impending breakthrough.

The work of Albers-Schönberg was critical to Forrester's success. The ceramics maker hit upon the right ingredients and proportions of ferric oxide, ferrous oxide, and manganese dioxide, which he shaped in a press, fired, and then cooled to form the cores. The "old German ceramicist, who by guess or by intuition, or years of experience, mixed up his first mixture and hit exactly the same spot as all the research done at MIT."

Some called them "magnetic torroids"; "Cheerios" is what Olsen aptly called them because they resembled the size and shape of the breakfast cereal.

The most important step came next—wiring them into an array, like a miniature chain-link fence with Cheerios strung throughout the links. "The core memory idea was not the idea of storing information in a core. That had been done before," Olsen said in an interview years later. "The clever idea of Jay Forrester's invention was the way of selecting the core." Forrester created an array of cores that were 16 by 16 equaling 256 cores, and through the center holes of each were threaded four sets of wires: one set of wires ran north to south, another set east to west, with a third and fourth set running diagonally through the holes.

Applying a selective current to any select wire made it "possible to choose any core within the solid array and to determine its existing state of magnetization or to establish it in either of its two desired states." So magnetized in either of two opposite directions (clockwise or counterclockwise), each core could then be used for storing a bit of information. This "coincident current memory" made it possible to address any specific core for reading or writing information into that core.

The two "desired states" were the equivalent of being "On" or "Off" or, in binary, a "0" or "1." "The ability to pinpoint specific intersections or addresses within the core rings, from which information could be stored and then recalled at random, created an unparalleled innovation in computing.

"The computer's central processing unit and its memory of stored data, procedures, and programs could now be operated interactively. Random access memory was born."

The Memory Test Computer first tested the 16-by-16 core plane. It worked. The next test would be on a 32-by-32 core arrangement for 1,024 cores. Another success. From April into June of 1953, a 32-by-32 core arrangement stacked 17 planes high—the anticipated memory size for installation in Whirlwind—worked to perfection.

On July 1, 1953, bright boy David Brown's group ordered 250,000 cores to be sent to IBM for the core memory fabrication and installation on the XD-1 and XD-2. For Whirlwind, its first bank of magnetic-core memory was wired in on August 8, 1953, the second on September 5, 1953 ... just weeks previous to running the Cape Cod System's first tests. Forrester had come through in a big way.

Guys like Doug Ross were ecstatic because memory now cycled at 9 versus 23 microseconds for electrostatic memory, and accomplished 40,000 operations per second instead of 20,000. "The memory never failed," remembers Forrester, which was a decided bright spot in an otherwise fragile new technology, "and Whirlwind had marginal checking to find drifting components before they caused trouble."

The bright boys, Project Lincoln, and the Air Force were ecstatic; IBM and the rest of the computer industry were jumping for joy as well, because they could then trash all the damn bottles forever. It was the breakthrough that the industry needed in order to move digital computing into complete primacy.

Valley's respect for Forrester and the bright boys was sky high. Magnetic-core memory, he said, "was a classic story of luck and pluck: a true epic out of the nineteenth century. The cores developed while I watched, and like a boy reading a Horatio Alger novel, I was inspired."

Of course, Valley did more than watch: he got the money—millions, he said—from the Air Force. The Air Force bet on a very long shot, and won. "Magnetic cores were simply not favored by the smart money," said Valley, "it was unlikely that perfecting them would have been regarded as commercially profitable."

Philips Laboratories, the vaunted Dutch high-tech giant and world leader in ferrite technology, called the idea of core memory an "American exaggeration." Even Bell Laboratories had shunned magnetic materials. Imagine the collective gulp they all took when Olsen cranked up the MTC and it purred, and when the bright boys trucked over 250,000 cores to IBM.

As great a year of discovery as 1953 was for Forrester, it was also a harrowing year where everything seemed time compressed, filled to the brim with breakneck activity, deadlines, tension, and heavy responsibility.

A false alarm of attack from Nunivak, Alaska (April 1952) had caused the country's first-ever nationwide Air Defense alert. The fallout in 1953 was that the guy who pulled the alarm, Air Defense commander General Benjamin Chidlaw, was over at the University of Michigan's Willow Run Facility ogling their air defense system. Whirlwind's competition was not much, but it had the attention of the powerful. And sometimes that's all an inferior idea needs in order to topple a great one. Willow Run caused a ripple of dissent to move through the Air Force, eddying up uncomfortable fears in both Valley and Forrester.

Other airmen, majors and colonels, who worked with and firmly believed in Whirlwind, sat Valley down for a talking-to late in 1952. The Barta Building was abuzz with important air defense activity, much of which was particularly confusing for these officers in charge. Between August 11 and August 30, Whirlwind was completely shut down and refitted with all the elaborate Input/Output hardware needed for the air defense system: viewing screens, operator consoles, Charactrons, light guns, magnetic drums, modems, telephone and radio communications, and all the other new accoutrements of tracking, scanning, and interception.

The officers wanted a blueprint, a document that totally visualized for them what this air defense direction center would look like, so that they could more fully envision the military aspects of their role in it all.

Valley could see the reasonableness and sincerity of their request, one which he and his engineers had totally overlooked. Valley admitted that he and his engineers had assumed too much of the officers "treating them all like General Jimmy Doolittle, who has a doctor's degree in aerodynamics."

Like the layout for a new home, the buyers wanted to see the architect's drawings and plans, and not make do with verbal descriptions and grand talk. Grand talk certainly would not do when reporting to the high command who would want to hear every detail—all the specifics, especially where the general's command chair would be placed.

Technical Memorandum 20 or just TM-20 was the result. When complete, they asked the recently retired Gordon Saville to critique their handiwork.

As Bob Wieser remembers it, Saville, all of "five and a half feet tall, feisty…and with a strong voice," strode to the head of a meeting room table and slammed TM-20 down. "You are," he bellowed, "the worst damn salesmen I ever met. This report is stinko profundo." Wieser recalls listening to Saville very carefully, and began "to understand that it's one thing to explain something that lies outside a person's experience and yet another thing to explain something that lies outside a person's imagination."

They rewrote TM-20…and Saville smiled his approval.

As necessary as TM-20 was for the Air Force officers, and as huge as was the impact it eventually had on the future training needs of the air defense system, Forrester needed, once again, to yank himself away from his pursuit of core memory and put together a how-to manual.

Officially it was titled, "A Proposal for Air Defense System Evolution"; unofficially it was a proposal for knitting together Air Force officers and engineers into a team.

Valley, Boehmer, Harrington, Forrester, Everett, and Wieser wrote the proposal and delivered it in January 1953. For Forrester it was just another distraction, albeit an important one, but nonetheless a time-consuming distraction in a year filled with distractions. The effort was worthwhile. TM-20 was widely distributed in the Air Force and made an impact.

In March, high-ranking Air Force Generals Partridge and Putt, the heavyweights of the Air Research Development Command, visited both Willow Run and the Barta Building, and on May 6 declared Whirlwind the champ.

Then, to add even more hurry-up in Forrester's rush to make Whirlwind ready, the Soviets added their own touch of anxiety. On August 12, 1953, they lit off the big one, a hydrogen bomb, the H-bomb, equaling 1 million tons of dynamite. Just one of these big bombs dropped anywhere had the destructive force of 70 Hiroshima-type atom bombs going off simultaneously, with a flash brighter than the Sun, and with a thermal flash of over 1,400°F (800°C) traveling at the speed of light.

The folks who got real jittery during the Nunivak scare, including the general population, suddenly saw not just mushroom clouds but apocalypse. The bright boys were center stage. Air defense was a must!

The bright boy most pleased with the success of Forrester's Cheerios was Bob Wieser, who was in charge of the Cape Cod direction center in the Barta Building. His big day on stage was to present to the Air Force a functionally complete experimental prototype of how an actual SAGE direction center would work.

In a world just recently introduced to digital computers, where human interaction with a computer was exclusively done by loading IBM cards into the machine and then waiting for an answer, Wieser would perform by conducting an orchestra of 50 manned workstations and monitors (what were then called consoles).

A cabinetmaking shop across the parking lot from the Barta Building built the consoles, and then Wieser and company crammed them full of electronics. Having two of Forrester's 1,024-core memory banks installed in Whirlwind seemed a godsend. The operator consoles had to be integrated by software, which, he said, was "the largest real-time control program ever coded, coding in machine language, since higher-order languages had not yet been developed."

Having solid memory to rely on was vital when simultaneously integrating hardware, software coding and debugging, and training Air Force personnel on the use of the equipment. When the September tests rolled around, Wieser was ready.

A long-range FPS-3 radar was installed near the tip of Cape Cod on a hill in the town of Truro, Massachusetts; gap-filler radars equipped with Harrington's "slowed-down video" went up to the south and north of Boston at Scituate and Rockport, Massachusetts. Each site also had an IFF (Identification Friend or Foe) system multiplexed to the radar information. Large spools of telephone cable dotted Route 6 leading to Truro, as well as the roadways leading to the other radar site—spools that would be unwound to connect up all of the far-flung sites.

The bright boys' modem, the John Harrington and Paul Rosen 1,800 bits-a-second modem that AT&T scoffed at as impossible, was a reality. "They said if one drives a telephone line over 600 bits per second, one has arrived at the end of the flat earth," recalled Rosen. Soon after, "the Bell Labs people adopted my modem" as the company standard.

Networked together with telephone lines and modems, the system was in readiness. All of the sites' data traveled over dedicated telephone circuits to the Barta Building, right into Wieser's console-crammed direction center. This setup was, in miniature, a test-bed for mock air attacks on Boston, a major U.S. city with a large population. If the system played well at protecting Boston, the model could be then extended out to cover all of North America.

For the Air Force, this command and control of air power using a computer and interactive consoles was the wave of the future. Amid the glowing confines of Room 222 with its Charactron screens, Area Discriminator, and indicator panels of dancing lights, the Air Force could see the air operations center of the future.

Wieser's room performed wonderfully well, and continued the performance as the radar network expanded south to New York at Montauk, northwest to Derry, New Hampshire, and as far north as Brunswick, Maine. SAC B-47s attacked Boston, while Air Force F-89s and F-86s, and Navy F-3s successfully intercepted each bomber in each formation.

By 1954, with the inclusion of automatic ground-to-air data links, Whirlwind was guiding interceptors to targets using the aircraft's autopilot, which was a prelude for future interceptions using pilotless BOMARC missiles. As Wieser said of his mates, "The workers were young, bright, enthusiastic, and very much aware that they were working on the leading edge of something new and important; they were learning on-the-job skills that schools did not teach. The hours were long, the camaraderie was close, and everyone wanted to make it work."

For Gordon Saville, this was the culmination of his entire career blooming out of an old brick building in Cambridge. Somewhere, Forrester must have been reflecting back a bit on the long journey that had begun back in 1947 with $L1$ and $L2$, when he and Everett announced that such a world would come to pass.

Back then, people brushed them off with a courteous smile. Not any longer. Herman Melville wrote, "It is better to fail in originality than to succeed in imitation." Maybe. But oh, how sweet it is to succeed in originality.

Chapter Eight
Voices in the Machine

Giant Brains, or Machines That Think

The XD-1 arrived at Lincoln Laboratory on January 5, 1955, and was assembled and running by January 12. Building F, a windowless, concrete, three-story blockhouse attached to Lincoln Laboratory, would be its home for the foreseeable future.

Building F was the model direction center for all that would follow it, complete with rooms for radar mapping, aircraft identification, weapons direction, battle station, and command post. XD-1, the new cookie-cutout Whirlwind, all spiffed up with 64 k of magnetic memory courtesy of IBM, and sleek looking, like a new Pontiac fresh off the assembly line, tucked its generous proportions easily into the cavernous citadel.

Somehow, during its transformation in Poughkeepsie from one-of-a-kind original to assembly-line knockoff, the machine had lost some of its charm. Switching on the XD-1 was a bit anticlimactic, more like switching off an era. The bright boys respected and admired the new creation, but there was something endearing about the old machine that the new one could not match.

"I felt that the pioneering days in computers were over," recalled Forrester in an interview 40 years later. Building the monster in the Barta Building was where the real fun had been. Things would never be quite the same again. From here on in, it would be perfecting the assembly of factory-made Q-7s for delivery and fulfilling the contract to network the direction centers across the continent. The real action,

Lincoln Laboratory (Building F blockhouse, center left), 1955. (Courtesy of MITRE Corporation.)

the zest for experimentation and discovery, was over when the last of the Cape Cod System tests concluded.

Forrester would be gone in a year. Valley left the very next year to be chief scientist for the Air Force. Everett checked out a year later. The following year, January 1959, the bright boys would split up forever. Some remained behind at Lincoln Laboratory to continue on in high-tech electronics, some left to seek new horizons elsewhere, and still others migrated a few miles away to join Everett at the newly formed MITRE Corporation to work the air defense contract for the Air Force.

The partings were mixed. Many times, as a technology matures and moves into wider acceptance, its originators and developers get chilled out of the flow of things. They lose control. They then move on or are moved out. Also, too long their own bosses, the aspect of heeling to new management and administrative masters was something some would never abide.

Plain to see for most was the fact that private enterprise and not universities was the future incubator of new computing machines. "I felt I had been in the field long enough," reasoned Forrester. Indeed, 10 years of building Whirlwind and managing the bright boys, plus the war years in the Servomechanism's Lab, was a long time. "I think you can argue that the ratio by which computers improved in the decade 1946–1956 was probably bigger than in any decade since."

Mulling over an offer from MIT's Sloan School of Management and nurturing a new direction for his talents, Forrester retired from the computer wars in June 1956.

Wildcats on their own for a very long time in Cambridge, it was inevitable and understandable that the new order of things would affect them. In 1954, *Time* reported on a survey conducted for *Fortune* magazine by Francis Bello that sought to find out "What kind of a man becomes an outstanding scientist." Far and away the most common "characteristic of outstanding young scientists [under 40]," he reported, "is a fierce independence. This is invariably coupled with a strong desire to work on the most crucial problems in their field."

For a long time the bright boys drank heartily from both of those wells. But as they said in their own brochure, "Digital computation is a new and growing field and a young and rapidly developing profession." Certainly, there must be other wells equally as sweet and heady as Whirlwind in that bright new world of digital computing out yonder. One never knows. The odds seemed in their favor and many went hunting for something few would find again.

As Wieser mused years later, "Sometimes I ask myself why this was such an interesting experience, the like of which I haven't had since."

The experience, much muted, continued on for the bright boys at Lincoln Laboratory. They were needed. They stayed a few years to finish what they had started. There was after all a national crisis at hand that demanded an air defense umbrella over North America. Certainly their patron, the Air Force, was watching in earnest, waiting for a solid return on its investment.

The bright boys could also take pride in the fact that they were now founding fathers, ascending from cocky wise guys to venerable pillars of digital computing. The acres of cinderblock walls and miles of black and white floor tiles that was the new laboratory were there because of them.

If Yankee Stadium was affectionately known as the house that Ruth built, then Lincoln Laboratory was the house that Whirlwind built. Without Whirlwind the Lexington woods would still be woods. And without the bright boys there would certainly be no Whirlwind.

Even the name Whirlwind was quickly receding from common parlance. In 1954, what formerly was Whirlwind, then the Cape Cod System or Lincoln Transition System, donned its official name: Semi-Automatic Ground Environment or SAGE.

At a press conference at Lincoln on January 16, 1956, SAGE was formally introduced to the public. And from that day forward, the machine that powered it all would be referred to as the SAGE computer.

With SAGE in the house, Lincoln grew rapidly. In terms of revenue it soon dwarfed its parent; its 2,000 staff, of which 700 were science or engineering professionals, pulled in $20 million a year. Digital computing saw dozens of electrical engineering research assistants working on their master's degrees while the staff sponsored 21 doctoral theses.

Soon MIT was whistling a happier tune about its air defense research lab 12 miles out of town. Lincoln had become an essential part of MIT's educational mission. As one of its directors asked rhetorically, "What can Lincoln Laboratory do for engineering education? It can provide entirely new opportunities to advanced students who wish to work in the complex fields at today's technological frontier." Today's technological frontier, no less.

The bright boys and their technology were suddenly a well-accepted part of the establishment. There was no denying the fact that Whirlwind's DNA was everywhere at Lincoln; it dominated the environment, and everyone had their hands deep within its genetic code rearranging bases into new forms. There was, of course, the XD-1, then TX-0, TX-2, the radars and modems and communications and peripherals. It set off chain reactions of ideas that skittered down every corridor and rattled the door of nearly every laboratory.

J. C. R. Licklider was a case in point. Wes Clark, the builder of the TX-2, first met Licklider one evening in the basement of Lincoln. "Lick," as he was known to his friends, was trying to operate a console. Clark briefly introduced him to the TX-2, its display screen and programming techniques. Licklider was hooked.

He'd return later, either alone or with others, for more pointers from Clark. "What Licklider gained from these subterranean sessions," wrote John Naughton in his *A Brief History of the Future*, "was a profound conviction about the importance of interactive computing... which he later articulated in 'Man–Computer Symbiosis'."

A machine was influencing a man's thinking; the man, in turn, would later use that new-found thinking to revolutionize the machine, and human interaction with it. This was the enduring legacy of the bright boys' tool placed into the hands of the right person at the right time.

The same TX-2 placed again into the hands of Ivan Sutherland brought forth a revolution in computer graphics with Sutherland's *Sketchpad* program (1963). Whirlwind's revolutionary idea of many consoles sharing the same computer, as in Room 222, would later galvanize young MIT researchers Fernando Corbato and Robert Fano into building a time-sharing computer system (1961). Time-sharing, they wrote, "can unite a group of investigators in a cooperative search for the solution to a common problem or it can serve as a community pool of knowledge and skill on which anyone can draw according to his needs."

Decades before the personal computer, computing was getting personal and social. In an age when people's only interaction with a computer was to stuff stacks of IBM cards up a computer's rear and then wait, Whirlwind's DNA was doing magic. For Licklider, his time with TX-2's screen was a profoundly moving experience. Interactivity became his drumbeat, a driving force of the most personal and social kind, heralding, as one observer describes it, "the crown jewel of Licklider's crusade... his initiation of events leading to the Internet."

(Detail) Building F blockhouse (center left), Lincoln Laboratory, 1955. (Courtesy of MITRE Corporation.)

In 1958, like a planetary probe heading into a new world, Clark sent the TX-0, slightly modified, winging off to MIT, because, as he put it, "... what MIT needed at the time was a computer that you could get your hands on. That was the one thing that they did not have." On the receiving end were the likes of Jack Dennis, who used TX-0 to build his MACRO assembler, and Tom Stockham, who built the FLIT debugging program, both of which were firsts of their kind in computing.

TX-0 transistorized computer at Lincoln Labs, 1956. (Reprinted with permission of MIT Lincoln Laboratory, Lexington, Massachusetts.)

Another MIT "Tixo" alum, J. Martin Graetz, summed up his time with TX-0 as, "the chance to work on this computer was in many ways a rite of passage; it meant that I had joined the ranks of the real programmers."

New Voices

In 1955, new voices were heard in the corridors of Lincoln Laboratory—eight new voices fresh in from the West Coast, from Santa Monica and the RAND Corporation. Eight computer programmers arrived to partner with the bright boys.

The XD-1 was scheduled to get a fresh new computer program, and RAND was hired to do the job. "The Air Force selected RAND as the only logical choice for this activity [RAND was created by Douglas Aircraft to do research exclusively for the Air Force]. One reason: RAND had a corner on the country's programmers."

There were no more than 200 programmers in the country and 10 percent were RAND employees. James Wong was one of them.

Wong graduated from UCLA in 1952, and went to work for RAND as an associate mathematician. RAND had an IBM 701 as well as its famous von Neumann clone, the Johnniac, for which Wong and Cecil Hastings developed programming for high-speed digital computers for RAND's System Development Division. RAND's job with XD-1 was to take over the operational program development, create supporting software, and then carry out the installation of that programming in all of the SAGE direction centers.

Wong's first time in the computer room was breathtaking and mind-boggling: "You're engulfed in there. You get a strange feeling that you are part of the computer, and the computer is part of you! It's like a member of the family."

He got acclimated by going around with the equipment team to check out long-range radars, gap-fillers, and the height finders. While hobnobbing around, he met pockets of other contractors who, like himself, were cutting their teeth on the giant computer. Bell Labs, RCA, AT&T, Burroughs, and IBM had crews similarly "engulfed" in XD-1. The RAND contract called for 25 programmers, if RAND could find that many to hire. Eventually, between direct hires and training, RAND amassed 700 programmers, 300 of whom went to Lexington.

The System Development Division got so bloated with people and revenue that RAND spun it off as the System Development Corporation in 1957.

Wong's first assignment was to write a test program for the ground-to-air data link, but his big test came when he got the assignment to lead his team in writing the machine's program executive control (PEC), which today is a computer's operating system. The PEC turned out to be 230,000 operating instructions. Wong's team did it all in assembly language, since there were no higher-level languages yet available. It took three weeks to complete.

"In those days we had one card per instruction," and one evening at midnight they gathered and fed all their IBM cards into XD-1, hoping mightily that there were no errors and that the machine would eat every one with gusto. All worked out well. However, if any bugs cropped up, IBM offered a bounty to find and fix the errors: "... nominal awards were like $50 and $100, and it went up to $1,000–$15,000."

The PEC was followed by utility and support programs that rung up an additional 870,000 instructions. In all, over 1 million lines of code went into what was then the largest program and largest programming job ever.

By the end of 1957, "the official handover of the software function from Lincoln to RAND was complete." In 1958, Wong and 70 other programmers went to McGuire AFB, New Jersey, where for two years the blockhouse and two Q-7 computers had awaited their coming.

By June, the McGuire SAGE site was online, and with 40 programmers SDC did the same at Stewart AFB in New York. "Subsequent sites went on-line about every two months," with programming proficiency getting to the point where only 15 SDC personnel were necessary to crank each machine into operation.

In 1998, at a lecture titled "Vigilance and Vacuum Tubes: The SAGE System 1956–1963," James Wong reminisced about those early days of his youth and the youthful profession of computer programming. For many, many years, he said, the word around the indus-try was: "It was first done in SAGE. SAGE was the real-time, command and control computer-based system with capability so advanced," he told the audience, "that 40 years later, today, some of that capability can still be called state of the art … things

Exterior of first SAGE installation, McGuire Air Force Base, Long Island, New York, 1958. (Courtesy of MITRE Corporation.)

like multiprocessing, real-time database management, distributed processing, time-sharing, interactive displays, networking—they were all there in SAGE."

Wong and his team did all coding without the benefit of high-level programming languages. They were too early for the likes of FORTRAN, ALGOL, COBOL, and BASIC. Binary was the XD-1's native language; its arithmetic unit gobbled up with great alacrity machine code like this string: 00101111001011011000101101001110110111.

But for Wong and his team, or any other programmer of the day, such strings of binary 0s and 1s were mind numbing. SAGE was programmed in machine language but had some shortcuts called "pseudo-instructions," what today are called "macros." Four such substitutes for binary were:

ETR for Perform an AND to the Accumulator with a mask the size and position of the item; POS for Shift the accumulator so that the least significant bit of the item is in the least significant bit of the accumulator; RES for Shift the accumulator left so that the least significant bit of the accumulator moves to the least significant bit position of the item; and, DEP for Deposit into the word containing the item from the proper bit positions of the accumulator.

This concept of "pseudo-instructions" was implemented with the SAGE Communication Pool (Comm Pool), which was the central data definition used to assemble programs. To add two items and store in a third, the following sequence was used:

```
• Move ITEM1 to accumulator
    CLA ITEM1
    ETR ITEM1
    POS ITEM1
```

```
• Save ITEM1
    STO TEMP
```

```
• Move ITEM2 to accumulator
    CLA ITEM2
    ETR ITEM2
    POS ITEM2
```

```
• Add the two items
    ADD TEMP
    RES ITEM3
    DEP ITEM3
```

Early code "pseudo-instructions" for SAGE computer. (Courtesy of MITRE Corporation.)

Such "pseudo-instructions" were less direct than the binary code that was SAGE's own language, but they were a lot more "programmer friendly" than 001 0111100101101100010110100011 10110111.

In 1958, Bell Lab's John Tukey gave it all a name: "software." During the next decade, the 1960s, software would take root and begin to flourish. A *Business Week* article published in 1964, "New Tool—New World," crowned software as computing's "new tool," and predicted:

> As more and more human abilities are transferred into the computer through programming and thereby become part of a new kind of library of skills, the effect may be equal in kind to the change that occurred when written language appeared. Writing freed mankind from total dependence on memory and permitted the accumulation and selection of effective knowledge. The computer program in turn accumulates and preserves skills.

There are enormous implications in this—for business, education, training, and future employment. There is not much point in having someone spend a great deal of time mastering the kind of skill that a computer can learn in a few seconds by having a program fed into it.

Over time, software would inherit the earth by making relative simplicity from complexity, by creating human understandable computer languages that freed people from coding in machine language. "You needed an internal program that was smart enough and fast enough," said Jules Schwartz in his *Development of JOVIAL*, "to translate a programmer's instruction into efficient machine code."

Kicking off the hit parade of programming languages to come were John Backus and his handpicked IBM team of programming misfits, who, after three years of working the problem, came up with the first permanent solution in April 1957 (Stan Augarten, *Bit by Bit*).

The question they pursued, was, as Backus put it: "Can a machine translate a sufficiently rich mathematical language into an efficiently economical program at a sufficiently low cost to make the whole affair feasible?" Their answer was FORTRAN, an acronym for "FORmula TRANslating System." FORTRAN was not perfect first time out of the chute, but it was a great beginning. FORTRAN enabled programmers to write programs 500 percent faster, while the execution efficiency of translating FORTRAN code into machine language and executing commands suffered a reduction of only 20 percent.

Interestingly, shorthand English and algebraic formulas—the same method that FORTRAN used—were first used on Whirlwind in January 1954. Backus, writing years later, recounts how J. Halcombe Laning, Jr. and Neal Zierler had the first algebraic compiler running on Whirlwind.

In the summer of 1954, Backus was at MIT for a demonstration that served to confirm for him the elegance and naturalness of using a concise mathematical language for FORTRAN. The impetus for developing FORTRAN, as it was with many such programming languages of the time, was simple pragmatism.

"FORTRAN did not really grow out of some brainstorm about the beauty of programming in mathematical notation; instead it began with the recognition of a basic problem of economics: programming and debugging costs already exceeded the cost of running a program, and as computers became faster and cheaper this imbalance would become more and more intolerable," wrote Backus. The door was open for programming to bubble and percolate throughout computing.

XD-1's brethren, the AN/FSQ-31, would get a language of its own called JOVIAL, but the language proved cumbersome and soon fell to the wayside. However, the breakthrough process that FORTRAN opened for others would be repeated with successes like COBOL, ALGOL, and BASIC.

Sometimes programming languages took years to perfect and then were built out still further by others. In the 1970s, Dennis Ritchie of Bell Laboratories developed the programming language called C; 10 years later, Bjarne Stroustrup, also at Bell Laboratories, sought to make writing good programs even easier and more pleasant for the individual programmer. Standing on the shoulders of C, he built a newer programming language called C++. C++, although based on C and retaining a great deal of its functionality, was actually a separate programming language.

Like their hardware pals, early programmers were an inventive lot. "The programmer had to be a resourceful inventor to adapt his problem to the idiosyncrasies of the computer: He had to fit his program and data into a tiny store, and overcome bizarre difficulties in getting information in and out of it, all the while using a limited and often peculiar set of instructions.

"He had to employ every trick he could think of to make a program run at a speed that would justify the large cost of running it. And he had to do all of this by his own ingenuity, for the only information he had was a problem and a machine manual."

If Wong and his programming mates needed any respite from the rigors of taming XD-1, they could nip off a mere 2 miles away for an evening's diversion in the chummy confines of the Lexington Theatre.

In 1957, the hot flick in town was the romantic comedy *Desk Set* starring Spencer Tracy and Katherine Hepburn. Strangely, right after the opening credits ran, there was some splashy billing for IBM, singling itself out for credit as the hardware provider and set designer for Spence and Kate's costar, a computer named EMERACK, the Electro-Magnetic Memory and Research Arithmetic Calculator, nicknamed Emmy.

Tracey as Richard Sumner is Emmy's inventor and an MIT-trained efficiency expert. Hepburn playing Bunny Watson runs the corporate library. Emmy has arrived to "help" Watson's staff do their work faster and better; and Sumner is there to "help" ease the integration of his machine with Watson's very leery coworkers.

Actually, Watson's company bought two of Sumner's "big brains," one for research and the other installed in the payroll department. During the machine's first hellos with the staff, Wong and his buddies would have glimpsed a dawning in the future relationship between man and machine: the IBM opening credit was a sales pitch and also a public relations message to position computers as work assistants and not inhuman job jackers.

And who better to costar with that kind of PR message than America's darlings, Spencer Tracy and Katherine Hepburn. IBM even allowed a humorous computer glitch to sneak into the script when the computer sends every employee a pink slip … even the boss. Computers were on the rise, automation was on the creep in America, and IBM was packaging a friendly smile on its product line.

That same year in New York City there were 200,000 elevator operators opening and closing elevator doors and whisking passengers up and down buildings large and small. But not for long. Within a single decade nearly all would be out of work from the kind of job that would never ever come again. Machines would be smoothing the leveling of elevators at every floor and machines would be warning passengers: "Watch your step."

If Wong was at the Lexington Theatre in October, the newsreel before the feature film would have been more eye popping than EMERACK. Inescapable was the shocking news of October 4. The Soviet Union blasted Sputnik into orbit around the Earth, which had huge implications for an air defense system trained only to look for manned bombers. It would be some years before anyone figured out how to make a portly nuclear bomb light enough and svelte enough to stick atop a missile. And even more tricky, there was the larger problem of building a guidance system accurate enough to control a missile's flight thousands of miles to hit its intended target.

As of 1961, the USSR had but six of these long-range R-7 missiles, whose radio-control flight accuracy was dubious at best. The R-7 took too long to fuel, its above ground launch facilities were large and vulnerable to attack, and it could be held on standby for only 24 hours before the propellant seals began to fail.

Bombers were still very much the only real way of dropping killer blows on American soil. But that didn't stop the panic in the streets. And it also did not stop more than a few long faces in Building F from looking up at XD-1 and wondering, what next. President Eisenhower went on television and radio to reassure the American people that America's space program was second to none.

A Japanese newspaper called Sputnik "a Pearl Harbor for American science" while the British prime minister declared, "Never has the threat of Soviet communism been so great." Two months after Sputnik, the Air Force got an Atlas rocket aloft; and on January 31, 1958, barely three months after the Soviet launch, the United States launched its first satellite, Explorer I.

But the thunder had already been stolen by Sputnik, and American science and engineering turned into one huge, hyper catchup program. And from kindergarten through to post-graduate studies, the American educational system was scrutinized, tinkered with, and pumped full of cash to outrace America's Cold War competition. Neil Armstrong's footprint on the Moon in 1969 was one of the direct results.

Although most of the Sputnik-inspired fear syndrome was unfounded, everyone remained in panic mode for much of the following decade.

On the other side of the fence from Lincoln Laboratory was the 1,100-acre Hanscom AFB, which was hopping with activity from 1951 onward. By October 1953, MEW Hill had been leveled to make way for new runways, two large hangars were built, a headquarters building—Building 1600—went up, as did a chapel, base housing, and the dedication of the Electronics Research Directorate and the Research Services Division.

Five months after XD-1 arrived, the AFCRL moved from Albany Street to Hanscom, joining up with aircraft from 6520th Test Support Wing.

When XD-1 was ready to take over the network from Whirlwind, and then when XD-1 relinquished the network to the SAGE continent-wide Direction Centers, Hanscom AFB would be ready for both. The Air Force was going bigtime for high-tech electronics and doing it in a hurry.

Much of the behind-the-scenes work in creating this forward-looking Air Force originated with Muir Fairchild and Gordon Saville, especially Saville. Just before his retirement in the spring of 1951, when he was Deputy Chief of Staff for Defense, Saville hired Ivan Getting "to track new scientific and technological developments, and connect these with Air Force strategy and operations."

To assist him as deputy, Getting named a former scientific liaison officer, Colonel Bernard Schriever (who would eventually build and command the country's entire ballistic missile program). Saville had previously named General Donald Putt to be Director of R&D, and Louis Ridenour as Chief Scientist. All of these appointments would have far-reaching impact on the Air Force, which, barely 10 years after its founding in 1947, had vaulted itself into prominence as the premier high-tech service in the U.S. military.

When Getting resigned to take an executive position with Raytheon, Col. Schriever replaced him as Assistant for Development Planning; then in 1954, promoted to General, Schriever undertook the Air Force's most important R&D job, the Atlas ICBM missile program.

To build the missile system, Schriever aped the systems engineering approach and techniques learned from Getting and SAGE. His systems integrator and Lincoln Laboratory-type facility would be Ramo–Woolridge, led by Simon Ramo and Dean Woolridge. The practice of "concurrency," whereby many manufacturers simultaneously made parts for a single project like the Atlas missile, was exactly the same as the relationship of SAGE's myriad manufacturers producing parts for the air defense system.

He would later codify it as the 375-Series of Systems Management regulations. Schriever even copied his famous weekly staff meetings called Black Sundays from the bright boys' brand of engineering management, continuing on a line of succession from the Valley era, Friday ADSEC meetings and from Forrester's weekly skull sessions first at the Digital Computer Laboratory and still later at Lincoln Laboratory.

Schriever, in selecting his own deputy, chose General Charles Terhune over General James McCormack, which set into motion two events critical to SAGE's future. McCormack, passed over for the Schriever position, retired from the Air Force to become an MIT vice president in charge of government relations. In 1958, it would be McCormack who eased SAGE out of a Lincoln Laboratory grown weary of SAGE's monopolizing presence, and who guided it to its next home, the MITRE Corporation.

In June 1960, Ivan Getting left his vice presidency at Raytheon, moving to Inglewood, California, to head up the Aerospace Corporation, a wholly owned Air Force subsidiary for advanced planning, technical evaluation, and systems engineering of Schriever's ballistic missile programs.

Aerospace Corporation took over much of the former Ramo–Woolridge relationship with Schriever.

General Terhune would soon turn up as the commander of the Electronic Systems Division (ESD) at Hanscom AFB, which by 1961 would fall completely under the sway of Schriever as part of the Air Force Systems Command.

The bright boys' fingerprints for systems engineering, management, and outsourcing were everywhere in the new Air Force. Axed by the Navy in February 1950 and scheduled for complete mothballing in June 1950 as recommended by the "Report on Electronic Digital Computers by the Consultants to the Chairman of the Research and Development Board," the bright boys and Whirlwind had provided huge return dividends to their Air Force benefactor.

In addition, both the Polaris missile program that put the Navy back into the national defense picture, and the NASA Apollo missions that put humans on the Moon, utilized Atlas-type systems engineering. Each became the success that it was because of Atlas and SAGE.

And if not for George Valley's telephone call to von Karman and Air Force money flooding the Barta Building, Schriever might not have systems engineered much of anything without hugely overspending in time and money. And with Sputnik hysteria in the offing, some fast, positive response would be absolutely necessary.

The Air Force did not gravitate to computing and high-tech electronics because it was clairvoyant, but rather out of necessity arising from being the Department of Defense's anointed defender of the nation's Cold War skies. To be such a defender required having jet bombers, supersonic interceptors, and a bristling array of ballistic, intercontinental, and surface-to-air missiles. And in turn, all of this demanded—for design, testing, manufacture, and operation—the accuracy, speed, and reliability of digital computers.

General Henry "Hap" Arnold (considered the founding father of the U.S. Air Force) and von Karman put the Air Force on the right course: what Arnold, as far back as 1938, had called "sowing the seeds" of advanced technology. Schriever was reaping the harvest of Arnold's sowing and at the same time sowing more seeds of his own.

High technology was becoming an integral and inseparable part of Air Force culture. For the bright boys, the Air Force afforded them "maximum latitude and flexibility in the interpretation of contracts...fullest

availability of military plans ... and reasonable freedom to manage projects," note Redmond and Smith in their *Project Whirlwind: A Case History
in Contemporary Technology*. That was an unprecedented luxury in building a system that had so much national security riding on it.

Bob Everett was surprised and grateful that the Air Force, whose reputation was riding on it all, was not terrified or more circumspect. "Those
of us who were designing SAGE believed in it, and I don't know how we
could have done the job if we didn't. I was amazed at the time and I'm still
amazed at the unflagging support of the Air Force. Truly remarkable."

Beginning in 1954, the year after the successful Cape Cod tests and the
same year that the name SAGE began to take hold, technology began to
transform the way the Air Force operated. Eisenhower was in office as of
January 1953, and he informed Congress of his New Look strategy, which
relied on dominance in nuclear weapons, strategic air power, and the doc-
trine of "massive retaliation" to any Soviet threat against the United States
and its NATO allies.

Curtis LeMay, SAC commander since 1948, began making substantive
changes to his forces to gear up for Eisenhower's New Look. Just as SAGE
intended to disperse to 23 Direction Centers around the country, blockhouses for which were well underway, LeMay in 1956 similarly dispersed
his SAC forces to 29 U.S. and 10 overseas locations. Many of his U.S. locations were curiously close to SAGE Direction Centers, sometimes elbow to
elbow.

Soviet bombardiers would most definitely relish the idea of hitting a
SAGE blockhouse and a SAC base with a single bomb. But there was wis-
dom in LeMay's plan. SAGE direction centers were worth their weight in
gold for their capabilities in gathering information, military information
that might well be the difference between a bomber force getting airborne
or being caught on the ground. The military had nothing remotely com-
parable to the SAGE network. Information meant time, and time would be
the difference maker between retaliation and obliteration.

LeMay had sweated out the Nunivak scare of 1952. He remembered
how both the Air Force and the event had ended up as sensational copy in
Flying Saucer Review. Time could not dull his recollection of how his SAC
forces were caught on the ground and how the Air Defense Command
had been criticized by the Pentagon for undue panic and overreaction.
He knew that timely information was the key to readiness. He knew that
Nunivak had laid bare the Air Force's information deficits of woefully

poor radio communication and slow, leased Teletype circuits. He dumped all the slow stuff in 1956; dug a huge, underground command post of reinforced concrete at SAC headquarters at Offutt AFB, Nebraska; and ordered in a primary network for the transmission of emergency action messages (EAMs) to his SAC bases, which later expanded into the 465L system that included one of Whirlwind's descendants, the AN/FSQ-31, plus a viewing system called a "Quadrajector."

Because it was near impossible to train airmen to react to massive Soviet bomber attacks—when there actually weren't any—RAND created for the Air Force the first computer program that simulated such bomber strikes. With SAGE's live radar shut off and hundreds of fake inputs fed into the Q-7 computer, the Air Force simulated reacting to a Soviet attack and practiced getting SAC forces airborne in a hurry.

LeMay early on realized the necessity of SAGE information. During the war, he had sided with the nonsense of unescorted bombers and high-altitude daylight bombing that the Kammhauber Line of flak towers and the Luftwaffe had flamed into near extinction. It was all too distinct for LeMay to recall the B-17 raids over Germany, Bloody Thursday, and the void of information at 30,000 feet.

He and Vandenberg were saber-rattling buddies from way back, but they were not fools. SAGE might be an air defense umbrella over the continent, but for SAC it was a weapon. It was a formidable information weapon that harkened its clarion call back through the ages, right to Sun Tzu's *The Art of War*—information wins wars!

LeMay was hell-bent to get his fair share and more of it, and he understood: "The survival of strategic aircraft on a given air base was related to the degree of alert practicable and the warning time available. With the Distant Early Warning (DEW) Line in operation against Soviet jet aircraft, LeMay counted on getting two hours' tactical warning time and believed that it would be possible to get something like 60 percent of his aircraft into the air in this time.

"Against Soviet ICBM attack, however, the zone of interior bases could count on only about a 15-minute tactical warning ..." Well before Soviet bomb bays opened, LeMay intended for his B-47s and B-52s to be long gone.

DEW line radar data would be flowing into SAGE direction centers and networked throughout the system ... including right into Offutt's big computer. Information about weather conditions and information from radars south of the DEW line, picket ships, Texas Towers, and early-warning air-craft would be fed to the blue-screened situation room where LeMay and his command staff could observe all the radar tracks from any incoming bogey. All of it in real time! SAGE, integrated with Nike and BOMARC missile sites, also provided real-time status reports on U.S. defensive weapons.

When Soviet bomber technology produced the faster, higher-flying Tupolev-16 Badger and the Tupolev-95 Bear, U.S. F-86s and F-89s were left flatfooted to offer chase or climb to ceilings like 46,000 feet and above. Saville's interceptor program produced the delta-wing F-102 "Deuce": the first supersonic military jet at Mach 1.2 with a ceiling of 55,000 feet and a thousand-mile range. Pilots depended on the information flow to their cockpit displays from the SAGE data link. Even better than the Deuce was the F-106 Delta Dart, called the "Six" by its pilots. The F-106 had an onboard digital computer that when hooked up with SAGE allowed SAGE to select and fire weapons as well as autopilot the craft.

Most pilots preferred to do their own trigger pulling. "All you had to do was select the armament, shoot and return to base," recalls William Neville, a former F-106 flyer. "The computer locked onto the inbound target and onto you. You saw everything on the tactical situation display."

The Canadian Air Force opted for the twin-engine Mach 1.7 F-101 Voodoo, also SAGE compatible. Both the F-106 and the Voodoo with AIM-4 missiles and nuclear-tipped Genie missiles, connected to data flows from SAGE, shut the door to enemy bombers coming over the Arctic Circle. LeMay remarked that his new *gladius informatiensis* had made "the fundamental concept of a coordinated air battle and a defense in depth a practical reality."

On June 28, 1958, the enlightened commander of 1,500 jet bombers came to McGuire AFB to honor his new information weapon. It was the opening of the New York Air Defense Sector, and the dedication of the first SAGE direction center.

"SAGE does not think," LeMay told the crowd at the ceremony. "It gathers and stores information and presents a picture on which man can act. SAGE does not nullify the need for well-trained and proficient personnel. It enables such personnel to do a better job." LeMay's understanding of the new tool was converging with Licklider's.

Information Gets Technology

From the Barta Building to Lincoln Laboratory to a national grid of 23 direction centers (and another at North Bay, Canada), the bright boys' technology was spreading. With a dozen commercial computer and electronics firms contracting on the SAGE project, it was only a matter of time until the technology was refined, repackaged, and sold as commercial products.

It was only a matter of time as well for thousands of newly trained *digerati* to leave these same contractors for opportunities elsewhere. System Development Corporation was one of many that witnessed the outflow.

Claude Baum, in his *The System Builders: The Story of SDC*, follows the SDC diaspora of SAGE-trained programmers as they migrated from SDC to start companies or meld into corporate programming staffs. Many drifted into then-quaint, backwater California towns such as Palo Alto and Mountain View, getting ready, in many cases, for guys like Robert Noyce to hit town with cool ideas about microprocessors.

Moving information into the arms of technology could not be accomplished without government intervention. Individuals, universities, and corporate research laboratories simply weren't enough. Bright boys' technology took lots of government cash and government faith. That trend would continue. An important difference post-1957 was that SAGE's success gave the government confidence in technology; it believed that technology could come to the rescue with solutions for great national problems.

Had XD-1, the SAGE network, and continental air defense become colossal failures, government reluctance to throw money at a potential technological breakthrough would have been more than obvious.

Faith in American technology to meet the threats posed by the 1957 tests of Sputnik and the Soviet Sapwood ICBM bolstered the government's resolve to let the money flow. With a worldwide recession in full bloom from April 1957 to August 1958, and U.S. unemployment rates skyrocketing, government spending was under a microscope.

Although faced with a financial crunch and the howlings of Democrats looking to sweep Republicans out in the 1960 elections, Eisenhower spent. The old general in him was eyeing worldwide events and not the West Wing. NASA, the National Aeronautics and Space Administration, was formed from the old ribs of NACA, the National Advisory Committee for Aeronautics; the U.S. Interstate Highway System was begun; and a little-known, specialty agency was formed called ARPA (the Advanced Research Projects Agency). ARPA would become the engine that pushed the bright boys' technology the final mile to modern Information Technology and the Internet.

Intercontinental ballistic missiles and anti-ballistic missiles were what was on Ike's mind at the forming of ARPA. Well before Sputnik and Sapwood, U.S. intelligence had been aware of and watching Soviet missile firings. A huge, top-secret radar at Samsun on the Black Sea coast of Turkey monitored Soviet intermediate-range and long-range test firings into the central Asian desert, and U-2 spy plane flights in June of 1956 photographed the facilities in Kazakhstan.

The White House recognized that a single, concerted Soviet effort in missile technology had accomplished much, whereas in the United States, the rivalry among its three services had diluted missile development. In his 1958 State of the Union speech, Eisenhower promised to bring a halt to the rivalry.

Just before the Special Committee on Space Technology was to meet in February 1958 to decide, among other things, the fate of NACA, basic missile research, and who among the services would be the heavyweight of the missile program, Eisenhower and his Secretary of Defense Neil McElroy initiated Department of Defense Directive 5105.15, establishing the Advanced Research Projects Agency.

The directive gave ARPA a deep reach into "the direction or performance of such advanced projects in the field of research and development as the Secretary of Defense shall, from time to time, designate by individual project or by category."

With the Air Force and its Atlas, Thor, and the Titan missiles; the Army with Jupiter; and the Navy with Polaris, Eisenhower looked to ARPA to reign in some semblance of control. If the idea of an anti-ballistic missile seemed feasible, what was then joked about as "trying to hit a bullet with a bullet," then Ike liked ARPA for that role as well.

Another of the charges from Directive 5105.15 was "to think independently of the rest of the military and to respond quickly and innovatively to national defense challenges." Bold projects that advanced America's defense-related technologies got precedence. And that is where computers slipped into the mix. A year later, NASA would scoop up all the missile business, leaving behind for ARPA the visionary, blue-sky technologies associated with computing.

Wes Clark's TX-2 was certainly blue-sky enough. His TX-2 became famous for "advanced graphic display research;" and the TX-0 that he and Bill Papian sent to MIT became "famous as the favorite tool of hackers in Building 26." Since all computers of the time displayed information via Teletype only, graphic displays were of particular interest to ARPA. The idea was how "SAGE-like displays might be adapted to many types of computers, not just the big ones used to monitor air defenses."

Sutherland's *Sketchpad* was just such a graphics system that could be of great interest to ARPA. In fact, *Sketchpad* was more than just a display. It could store visual patterns as easily as alphanumeric data; and as such was a "simulation language that enabled computers to translate abstractions into perceptually concrete forms." The Building 26 hackers would qualify their blue-sky credentials when they used their TX-0 to become the core group who pioneered time-sharing.

Display graphics and time-sharing were two other large helpings of bright boys' technology poised to get exposure in the greater research community, joining keyboard input, magnetic memory, and the modem as essential bright boys' originals going public.

Of course, the real must-have companion for graphical input/output and time-sharing was Whirlwind's specialty: real-time computing—the linchpin for making the flow of information really worthwhile. No one was going to wait for information to meander to a screen. The essence of a screen was to see it now, which meant to see it fast.

In 1957, two months before Sputnik, the National Office Management Association conducted a survey of almost 4,000 U.S. firms with 5,000 office workers or more. The survey found that 50 percent had already installed a mainframe computer (each with a purchase price of over a million dollars), while another 14 percent were waiting for the delivery of theirs.

The vast majority of the work asked of these computers was payroll. These 2,500 companies, with an aggregate of 12.5 million workers, were spending $2.5 billion to automate only their payrolls.

Hollywood was correct: business looked exactly like *Desk Set*. The aspect of time-shared, real-time computing and graphical displays entering *Desk Set* environments all over America would be nothing short of revolutionary. These new tools, if they could edge their way in, would change the way information was processed, and as Licklider had predicted, they would "also change the way people thought."

Importantly, businesses had taken serious notice of computing and were investing in the technology. However, the going was slow. Grappling with card-fed computers was awkward. "Computers worked like machines on an assembly line—repeating the same operations as records were passed through them one at a time." An application like payroll might well be split into 10 or 20 runs, for example, deducting union dues from an already calculated payroll required a separate run. It was far faster and more efficient than payrolls worked by hand in rooms full of office workers with calculators, but it was still slow and cumbersome. This was a long way from the man/machine interaction already at hand with SAGE.

Needed as well as new computing tools was the realization that there was a "relatedness" to information the parts of which could be brought together to form a picture of overall business operations. As in LeMay's computer battlescape, where a computer "gathers and stores information and presents a picture on which man can act," so too could computers gather together information to present a picture upon which a business could act.

Business had no Barta Building in which to incubate and test; it would have to wait, borrow, and adapt both equipment and theory as the computers worked their way slowly into each business operation.

The dawning came slowly and in waves. "In 1953, during what was probably the first extended discussion of information as an abstract quantity to reach a large executive audience, Fortune magazine lauded it as a great and almost unknown scientific theory whose impact on society was likely to exceed that of nuclear physics." The article, "The Information Theory" by Francis Bello, focused on the technical and electronic communications aspects of Shannon's new theory of 1948.

New terms like "Automation," "Electronic Data Processing," and "Management Information Systems" began cropping up in management texts and journals, each with its own coterie of theorists and consultants pushing the new ideas into the workplace.

Howard Levin's 1956 *Office Work and Automation* called for a new breed of information specialists—information engineers—as well as a vice president for information to improve and refine the effectiveness of analyzing and using information. Putting together computers, information, automation, and management, two University of Chicago Business School professors in 1958 came up with the term "Information Technology." Their article in *Harvard Business Review* titled "Management in the 1980s" predicted a future workplace with corporations transformed by computer hardware, operations research methods, and simulation programs.

By far the most inspirational and clearly articulated look at computers was delivered by Edmund Berkeley in his 1949 national bestseller *Giant Brains, or Machines That Think.* Berkeley was the first to introduce electronic computers and their potential use in business to a general audience.

Thomas Haigh's paper "Lost in Translation" marvelously sets out the impact of *Giant Brains*: "Berkeley ... gave early expression to the idea of information as a ubiquitous presence in the natural and social worlds. He made the computer less threatening by presenting it as the latest and most powerful in a series of pieces of 'physical equipment for handling information' that included everything from nerve cells to human gestures."

The article "Do-It-Yourself Giant Brains" by Andrew Leonard in *Salon* hones in more closely on Berkeley's own well-tempered brain: "In language that managed the delicate trick of being exquisitely clear and uncompromisingly evangelistic, Berkeley described how a computer works, step by step, instruction by instruction. Employing numerous diagrams, and painstakingly explaining every underlying concept (like 'binary' or 'register' or 'input/output') as if it had never been explained before, Berkeley demonstrated how it was possible to move digital information from one 'place' to another—and how a set of on/off switches, if wired correctly, could perform operations on that information, handling such extraordinary feats as the addition of two plus two."

In 1947, while working for the Prudential Life Insurance Company, Berkeley was the guy who engineered his company's first computer purchase: a UNIVAC, which Eckert and Mauchly were unable to deliver before their company's collapse.

Drinking in every delicious page of Berkeley was a 25 year old named Doug Engelbart. Five years out of the Navy, directionless, drifting in and out of jobs on the periphery of computing, Engelbart was the exact right guy to profit from the Moore School lectures, but never went.

If the bright boys had a spiritual and intellectual lost member of the gang, that unique spirit was Doug Engelbart, future creator of both the modern concept of the graphical user interface (GUI) and the GUI's illustrious sidekick, the mouse. Engelbart was more than just an inventor of "things" that made computing easier and more intuitive; he wanted to use computers to help people to learn.

In *Tools for Thought*, Howard Rheingold devotes a chapter to this curious man's journey to recognition, titled "The Loneliness of a Long-Distance Thinker." "Loneliness" and "long-distance thinker" are key words for a young man with ideas aplenty stuck on the West Coast where he was looked upon as a crackpot by hostile colleagues.

At 211 Massachusetts Avenue, Engelbart would have fit right in. As Douglas Ross had before him, Engelbart would have gravitated to Room 222 for the epiphany of a lifetime. He would have looked at Bob Everett's light gun and seen the makings of his mouse; he would have seen in Bob Wieser's time-shared consoles his learning system for computer interaction with people; he would have seen in Forrester's dogged pursuit of magnetic RAM his own stubborn will to succeed against all odds; and, with the camaraderie of the bright boys for support, he too would have smiled at the vultures circling above the Barta Building. "I confess that I am a dreamer," said Engelbart years later. "Someone once called me just a dreamer. That offended me, the just part; being a real dreamer is hard work. It really gets hard when you start believing your dreams."

In 1951 in San Francisco, no one believed in this Oregon farm boy and World War II radar operator's vision of the close interaction of computers with people.

In Sputnik-laden 1957, Engelbart got a job in Menlo Park, California, at the Stanford Research Institute where research was ongoing into the scientific, military, and commercial applications of computers. It seemed a perfect match. But strangely and sadly for Engelbart, his ideas on computers interacting with people to "augment their intellect" did not fit in at all.

Rheingold recreated a telling Engelbart incident in his *Tools for Thought*: "How many people have you already told?" questioned a colleague after hearing Engelbart's theories. "You're the first," he said in reply. "Good, now don't tell anyone else," came the surprising rejoinder. "It sounds too crazy. It will prejudice people against you."

Curtis LeMay would have had full belief in Engelbart. But how was Engelbart to know that? Fortunately, the Air Force found him first. The Air Force Office of Scientific Research gave him the go-ahead and a small grant. The grant offered him the opportunity to freely explore the field. "It was lonely work, not having anybody to bounce these ideas off, but finally I got it written down." What Engelbart "got written down" for Air Force contract AF49 (638)-1024 was a remarkable, 139-page paper titled, "Augmenting Human Intellect: A Conceptual Framework."

This manifesto for "augmenting man's intellect" with "high-powered electronic aids" would lead to the development of computer-based technologies for manipulating information; in short, Engelbart was putting the "personal" into "personal computing." Upon its publication, ARPA took close notice, and soon Engelbart found himself pulled into the protective ARPA force field set up by Licklider and Wes Clark.

Engelbart read Berkeley's book in 1950 but did not produce *Augmenting Human Intellect: A Conceptual Framework* until 1962. That's 12 years of laboring in the vineyards before his salvation arrived. That's a long time to wait. However, if the event chain leading from Whirlwind to the Barta Building, to George Valley, to the Air Force, and then to ARPA had not happened, Engelbart and many others may well have labored obscure and anonymous for a very long time. Engelbart would indeed labor on for years more, but not in anonymity.

The first glimmer of Information Technology and its rise to prominence in the business world came in 1961 as IBM's SABRE system (SAGE's business-world alter ego) pushed to completion. The brainchild of Perry Crawford, SABRE was first pitched to American Airlines in September 1957, at which time the airline was quick to see both the critical importance to reservation bookings as well as the financial windfall that would usher up from such control.

Prior to SABRE, airline reservations were made manually in huge rooms by hundreds of workers transferring teletype information to large reservation boards at the fronts of the rooms. "Some of these rooms were so large that employees near the back would use field glasses to see the boards showing the number of seats available for sale."

SABRE was a real-time, time-shared system where travel agents at consoles individually probed the central computer in order to view airline flight schedules and to book reservations. It was like a civilian-style Room 222. James Gallagher's 1961 *Management Information Systems and the Computer* makes a case study of the SABRE system that "has been a textbook example of the strategic use of computers ever since."

And that was just the beginning. "During the mid-60s, computer makers cemented their commitment to the new vision of real-time, online, managerially oriented systems." Behind the façade of this new vision, people and technology would move from mainframe computer to minicomputer to personal computing.

The bright boys had poked a hole into the dome of Information Technology, and the stars of IT came tumbling down. For the bright boys, their job was over; it was for others to crack ever-larger holes in the dome. Engelbart would be but one of many.

ARPA in 1962 would go one step further. Under its then director Jack Ruina, ARPA hired Licklider to run its newly formed Information Processing Techniques Office (IPTO) to "extend the research carried out into computerization of air defense by the SAGE program to other military command and control systems."

In particular, IPTO would refine SAGE's wide-area computer network to build a survivable electronic network to interconnect three vital Department of Defense sites: the Pentagon, Cheyenne Mountain in Colorado, and LeMay's SAC headquarters. One of the IPTO research grants produced the memorable and important Larry Roberts TX-2 to Tom Merrill FSQ-32 digital packet exchange over dial-up telephone lines in 1965, which triggered the beginning of ARPANET, which later evolved into the Internet.

SAGE, as a defender of the skies over North America, had and still has its share of detractors. Some critics labeled SAGE as extravagantly expensive and a failure, rendered obsolete by Soviet ICBMs before a single SAGE direction center ever went into action. Closer scrutiny proves otherwise.

ICBMs did not become the threat of choice overnight; it would take time before bombers would play second fiddle to missiles. And very quickly did SAGE descendants become the eyes used to search for those ICBM launchings continents away.

Every modern military, friend or foe, uses a command and control system directly descending from the SAGE lineage. Every modern air traffic control system guides and lands commercial aircraft using methods pioneered by SAGE.

Today's computing business owes a huge debt to SAGE innovations, and should be especially grateful for the billions of dollars of revenue scuttling throughout the electronics industry because of SAGE ingenuity. SAGE was actually quite a bargain; it was peerless in buying peace of mind for millions of people and their governments.

For a few pennies a day per person, every man, woman, and child in the United States and Canada, not to mention much of Western Europe as well, depended on this bargain-basement insurance policy. It kept adversaries unsure and off balance, which was certainly far cheaper than a war.

Although none exists, one would half expect some tribute to be in evidence somewhere for the bright boys. Maybe a large bronze somewhere along the Charles River where Massachusetts Avenue meets the banks, maybe something with all the boys cavorting together. Something the sun could glint over and play upon showing off spirited young men forever young.

Ret. Adm. Edward L. Cochrane (MIT's vice president for industrial and government relations), George Valley, Maj. Gen. Raymond C. Maude, and Col. Dorr Newton at a press conference announcing the SAGE system for continental air defense. (Courtesy of MITRE Corporation.)

Cocky looks with wry smiles of confidence from youthful faces who knew that they were working on the biggest of big jobs and that everyone depended on them. Maybe a life-size chunk of glittering statuary memorializing their great challenge, their intellectual curiosity, their courage, their bravado.

Maybe something that is inspirational as well, something that a future Doug Engelbart in passing by might take heart from. But, sadly, no dice.

The bright boys have disappeared, seemingly erased. And their old haunt, the Barta Building, is still not a national historic landmark with commemorative plaque and slick visitors' brochure. Known on campus as Building N42, thousands of MIT students, most likely toting a digital device or two, passed through its front door unaware of the building's importance. In a slight nod of distinction, the IEEE in 2012 put a small plaque aside the front door.

Whirlwind was carted away in 1959 by bright boy programmer Bill Wolf for use in his own consulting company, Wolf Research and Development. Decades later, pieces of the great machine turned up at computer museums and at the Smithsonian.

Forrester and Everett turned up in 1989 at the White House for some long-overdue kudos from George Bush, the elder. Each received the National Medal of Technology some 40 years after the fact.

The only remaining witnesses at 211 Massachusetts Avenue are the gargoyles that encircle the building. They've seen it all.

And just maybe the best tribute to the bright boys are the rumblings 30 feet beneath Massachusetts Avenue as the Red Line subway clatters in and out of Kendall Square station every seven minutes. It was the bright boys who showed the world that computers could do more than just count. Every seven minutes in Kendall Square station the proof rolls in and out again.

Index